THE COMPLETE? POLYGON

BEING THE FIRST PART
OF SEVERAL COMPRISING
THE COMPLETE? POLYHEDRA
BY PATRICK TAYLOR

THE COMPLETE? POLYGON

First published February 1997 by

Nattygrafix
277 Cavendish Street
Ipswich
Suffolk
IP3 8BQ

(01473) 289861

Made in England

ISBN 0 9516701 2 3

NATTYGRAFIX IPSWICH

CONTENTS

ILLUSTRATIONS

For Tommy Stratton who saw the first spark

"When a long established system is attacked, it usually happens that the attack begins only at a single point, where the weakness of the established doctrine is peculiarly evident."

Bertrand Russell *An Essay on the Foundations of Geometry* 1897

INTRODUCTION

The material in this book is all about polygons, more specifically regular, star and cross polygons and their more important properties. Whilst not 'complete' in dealing with all properties of all polygons, it does cover those necessary to an understanding of the polyhedra and other assemblies of regular polygons in two, three or more dimensions. The Complete?ness is there in allowing any collection of equal length lines in the plane, meeting at equal angles and inscribed within a single circumcircle, to be described and classified by reference to a single fractional reference number. From this single label the properties of the polygon (angles, vertices, duals, area, truncation properties etc.) can all be deduced.

No apologies are made for the unusual approach taken to the subject. Essentially pragmatic, it has already borne fruit in the form of the author's 'Additions to the Uniform Polyhedra', where cross and double polygons are successfully used to generate some new and some previously missing or overlooked polyhedra. The approach works.

Having laid the foundations herein, the author intends to produce further volumes on an annual basis, dealing in turn with tilings, polyhedra, close-packs, skew polyhedra and hypersolids in approximate order of increasing complexity. At the end of the day he hopes to have written the larger work he would have liked to have purchased at the outset, when his interest was first stimulated by the wonders of symmetry.

REGULAR POLYGONS

A regular polygon is a plane convex figure composed of equal edges with equal angles between each adjacent pair, i.e. it is both equilateral and equiangular.

A k-sided regular polygon can be generated by adding and rotating a vector k times through an angle $\theta = 360/k°$. See Figure 1.

In polar co-ordinates θ is the external angle measured in an anticlockwise positive direction, which when applied k times ensures a full rotation through 360°, thus returning to the starting point.

The internal angle ϕ, that between the new vector and the old one, is therefore $(180 - \theta)°$, again measured positively in an anticlockwise direction.

Thus $\theta = 360/k°$ $\quad \phi = 180 - 360/k$

$$\begin{aligned} &= 180(1 - 2/k) \\ &= ((k - 2)/k)180° \end{aligned}$$

Consequently the sum of the internal angles of a k-sided polygon

$$\begin{aligned} \Sigma\phi &= k((k - 2)/k)180 \\ &= (k - 2)180° \\ &= (2k - 4) \text{ right angles,} \end{aligned}$$

the formula known to many which yields 180° for a triangle, 360° for a quadrilateral, 540° for a pentagon etc.

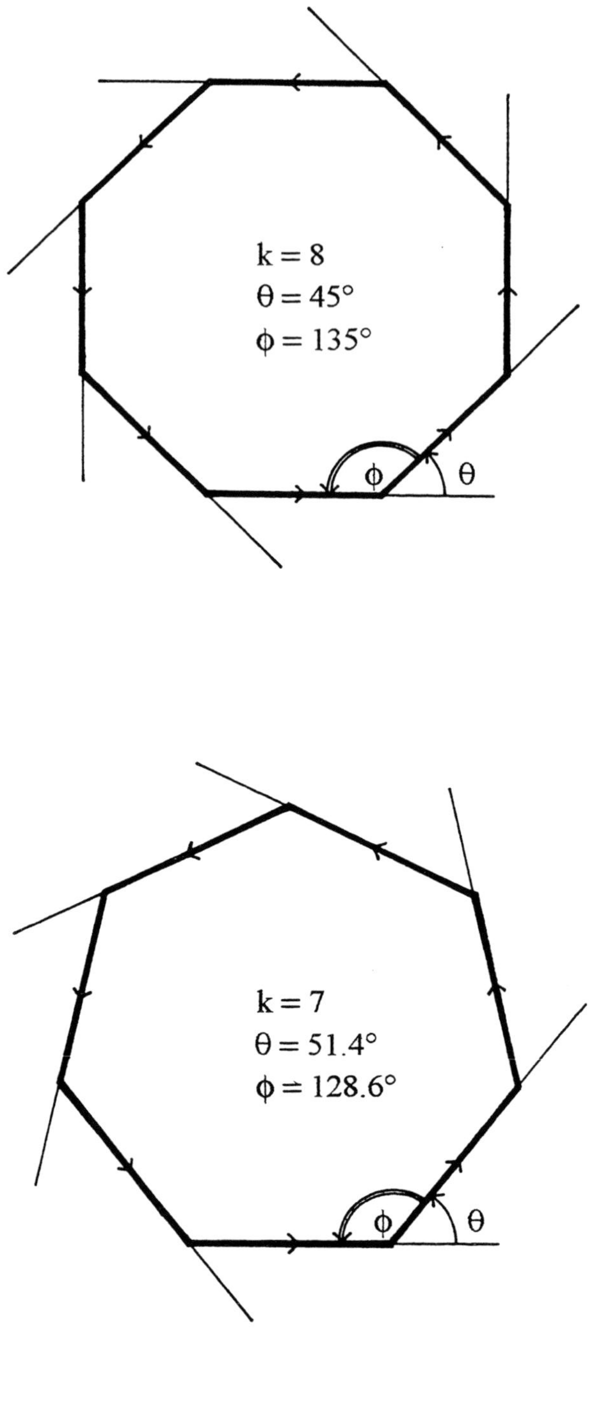

Fig. 1: Generation of Regular Polygons

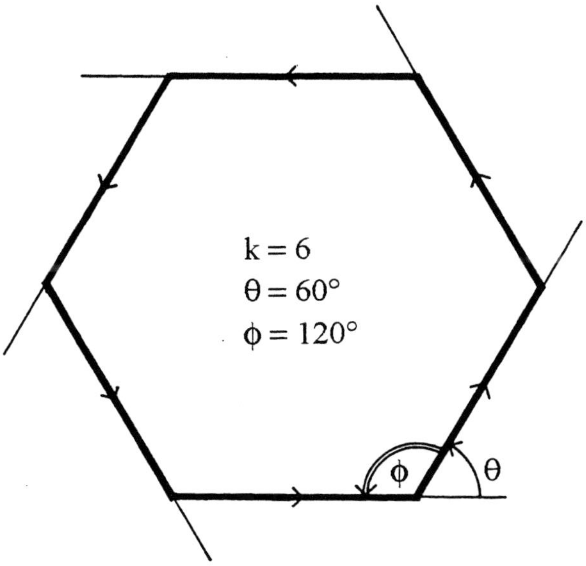
k = 6
θ = 60°
φ = 120°
φ
θ

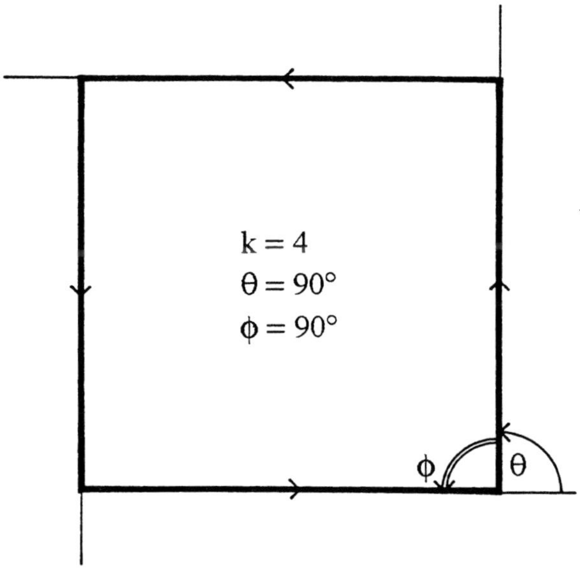
k = 4
θ = 90°
φ = 90°
φ
θ

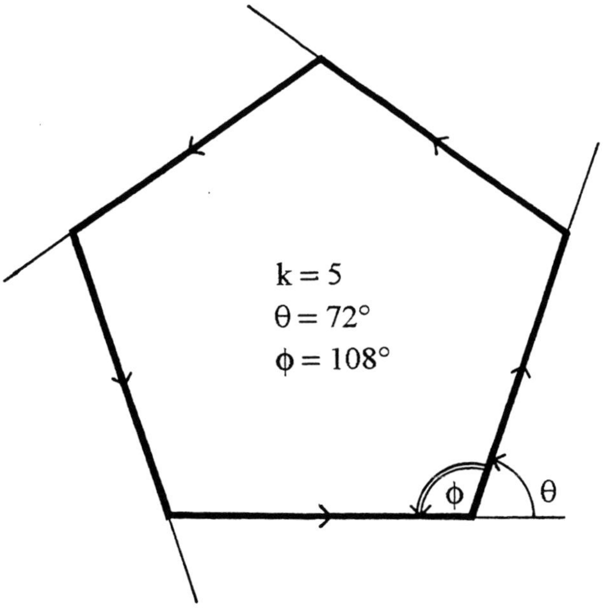
k = 5
θ = 72°
φ = 108°
φ
θ

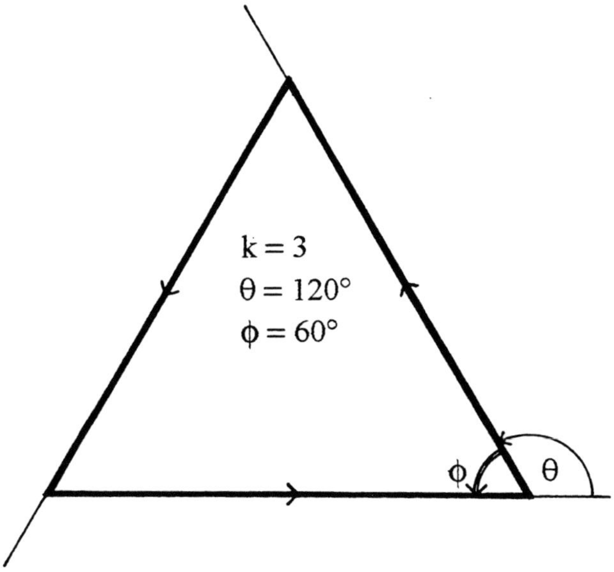
k = 3
θ = 120°
φ = 60°
φ
θ

MINIMAL POLYGONS

Interpreting θ as a function of k will yield a hyperbolic graph, but for regular polygons only integer values of k greater than 2 will have any immediately recognisable meaning.

On the graph of this function shown in Figure 4 the regular polygons can be seen to continue by integer leaps towards infinity, at which asymptotic point the polygon becomes a circle, $k = \infty$, $\theta = 0°$, $\phi = 180°$.

The regular polygons produced so far are summarised in the upper section of Table 1, but we might well ask what happens for other values of k? Do the values 2, 1 or 0 have any meaning as geometrical figures?

The lower section of Table 1 continues by looking at these values of k and using the formulae assigns values of θ and ϕ, which can then be interpreted geometrically.

Firstly in the case of $k = 2$ we get a digon, a polygon closing on itself with no internal area, the two vectors comprising it being equal but in opposite directions.

The value $k = 1$ yields a situation analogous to our infinite circle, but here the single vector has to close on itself, becoming infinitely small in order to do so. The polygon is effectively a point.

The value $k = 0$ is even more obscure, asymptotic like the circle, yet having no sides and seemingly infinite angles; it can only be regarded as no polygon at all.

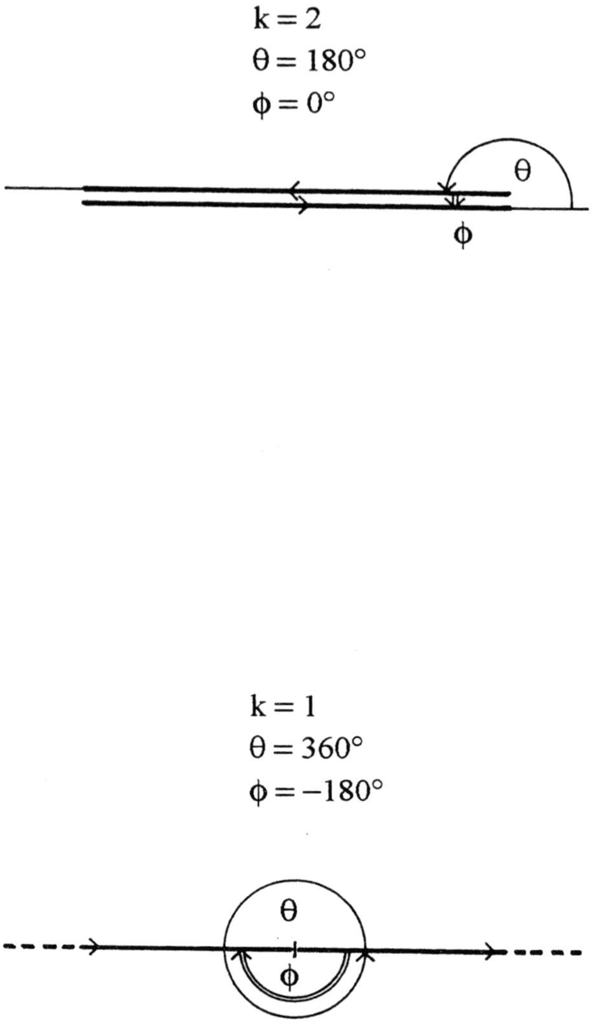

Fig.2: Generation of Minimal Polygons

	k	θ	φ		Polygon
	∞	0	180	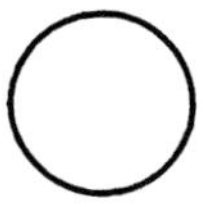	Circle
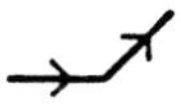	8	45	135	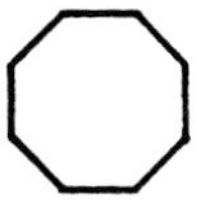	Octagon
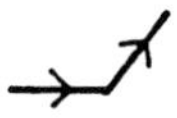	7	51.4	128.6	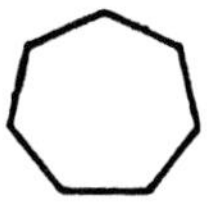	Heptagon
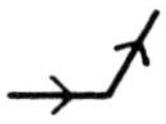	6	60	120	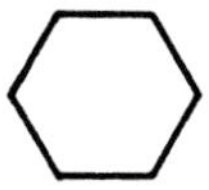	Hexagon
	5	72	108	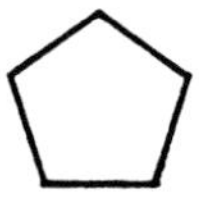	Pentagon
	4	90	90		Square
	3	120	60		Triangle
	2	180	0		Digon
	1	360	−180		Point
	0	∞	∞		O

Table 1: The Regular Polygons

NEGATIVE POLYGONS

Having now derived polygons for all positive integer values of k, we can now ask what happens for negative integer values?

Using the same formulae we can derive the results given in Table 2, where it can be seen that we get 'negative' polygons, with their vectors going around clockwise, rather than the anti-clockwise of the positive polygons.

One way to interpret these is by subtracting 360° from ϕ to make the internal angles negative ($k = -3$, $\theta = -120°$, $\phi' = -60°$). They then become exact mirror images of the positive polygons, a 180° turn out of the plane through the third dimension transforming one to the other.

Alternatively we can simply regard them as having an internal angle occurring outside the polygon ($k = -3$, $\theta = -120°$, $\phi = 300°$) The problem here is determining on which side of the edges the polygon actually lies. This can be illustrated by cutting a positive polygon out from a sheet of card leaving its negative behind, a kind of inside out polygon. If two dimensional space were actually spherical then either piece of card would be a valid polygon.

This is somewhat akin to the concept of electricity, where the flow of negative electrons is matched by an equal and opposite flow of vacated positive 'holes'.

There is thus a negative polygon for every positive one and they do not therefore warrant further separate investigation.

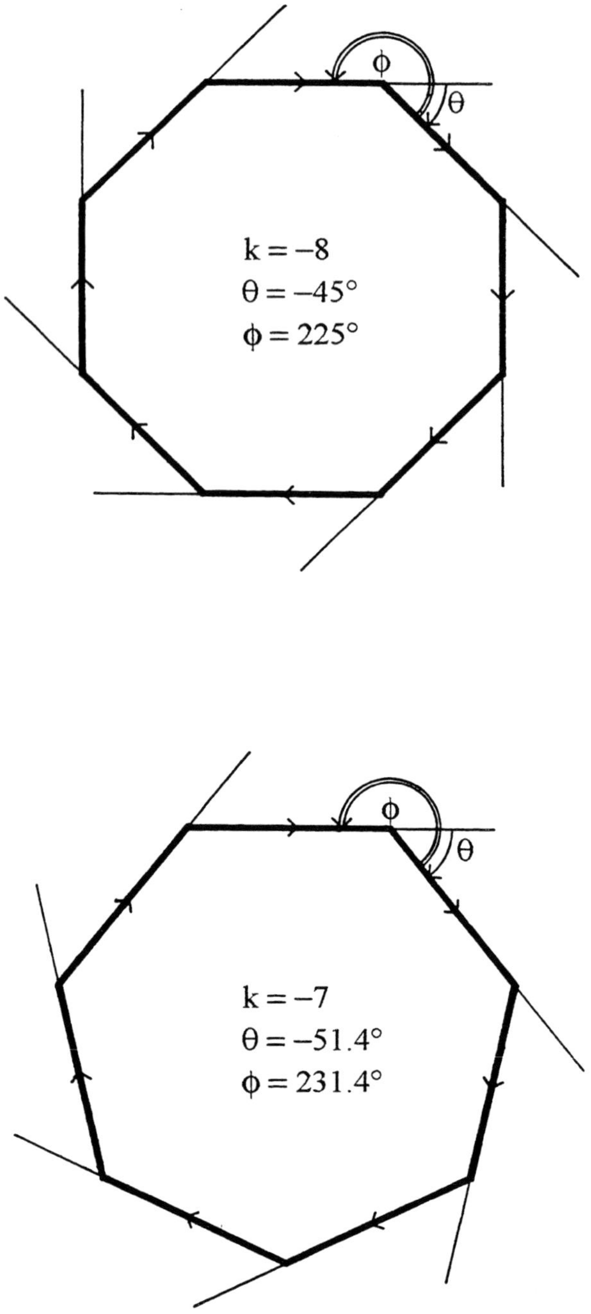

Fig.3: Generation of Negative Polygons

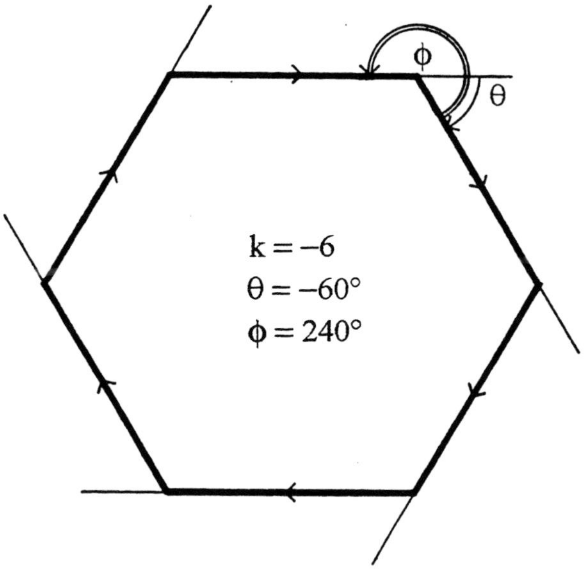
φ
θ
k = −6
θ = −60°
φ = 240°

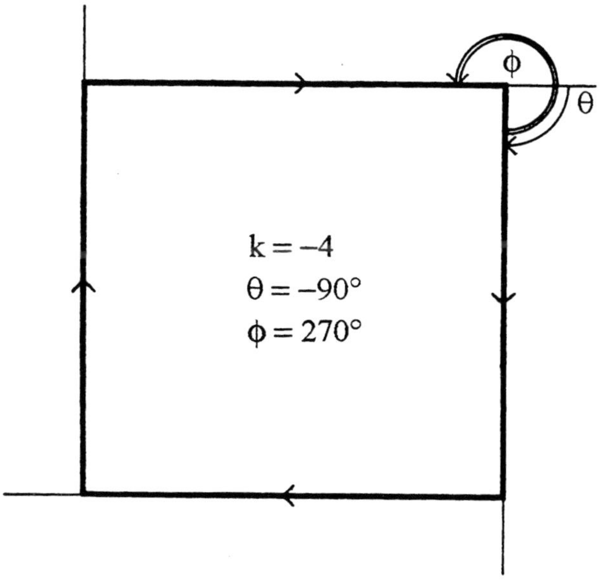
φ
θ
k = −4
θ = −90°
φ = 270°

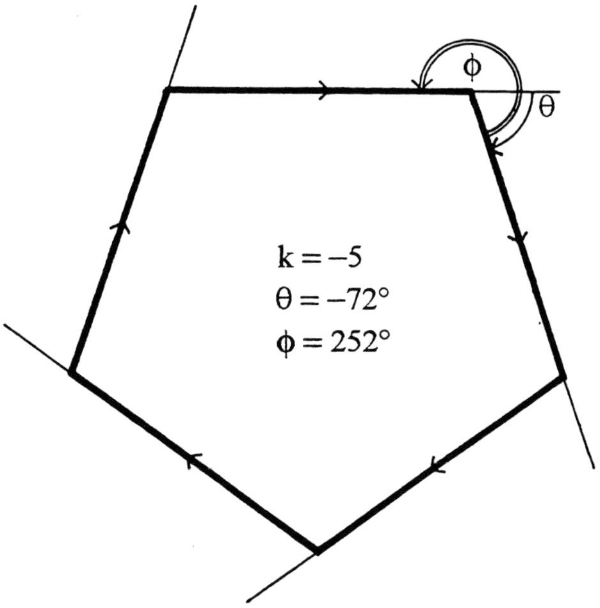
φ
θ
k = −5
θ = −72°
φ = 252°

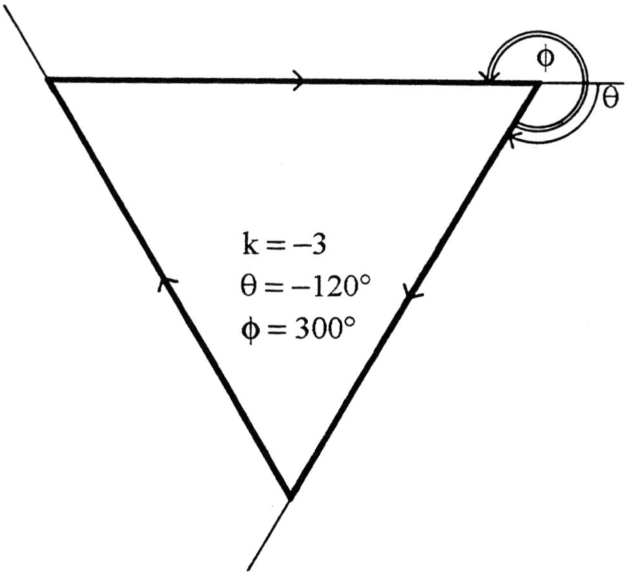
φ
θ
k = −3
θ = −120°
φ = 300°

	k	θ	ϕ		Polygon
	0	∞	∞		O
	−1	−360	540		Point
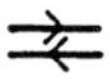	−2	−180	360		Digon
	−3	−120	300	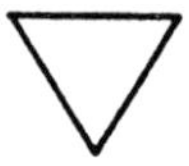	Triangle
	−4	−90	270		Square
	−5	−72	252	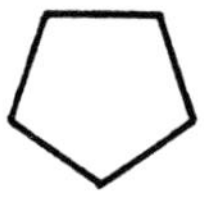	Pentagon
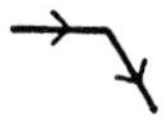	−6	−60	240	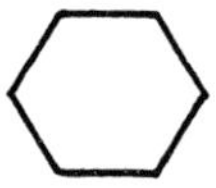	Hexagon
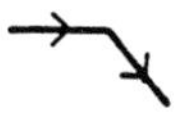	−7	−51.4	231.4	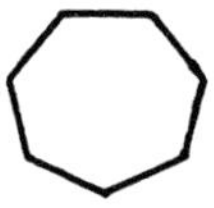	Heptagon
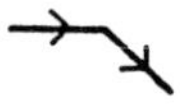	−8	−45	225	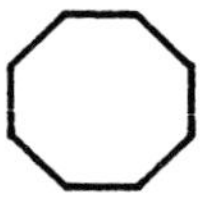	Octagon
	$-\infty$	0	180	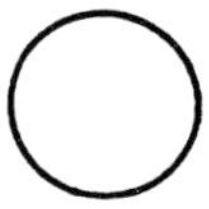	Circle

Table 2: The Negative Polygons

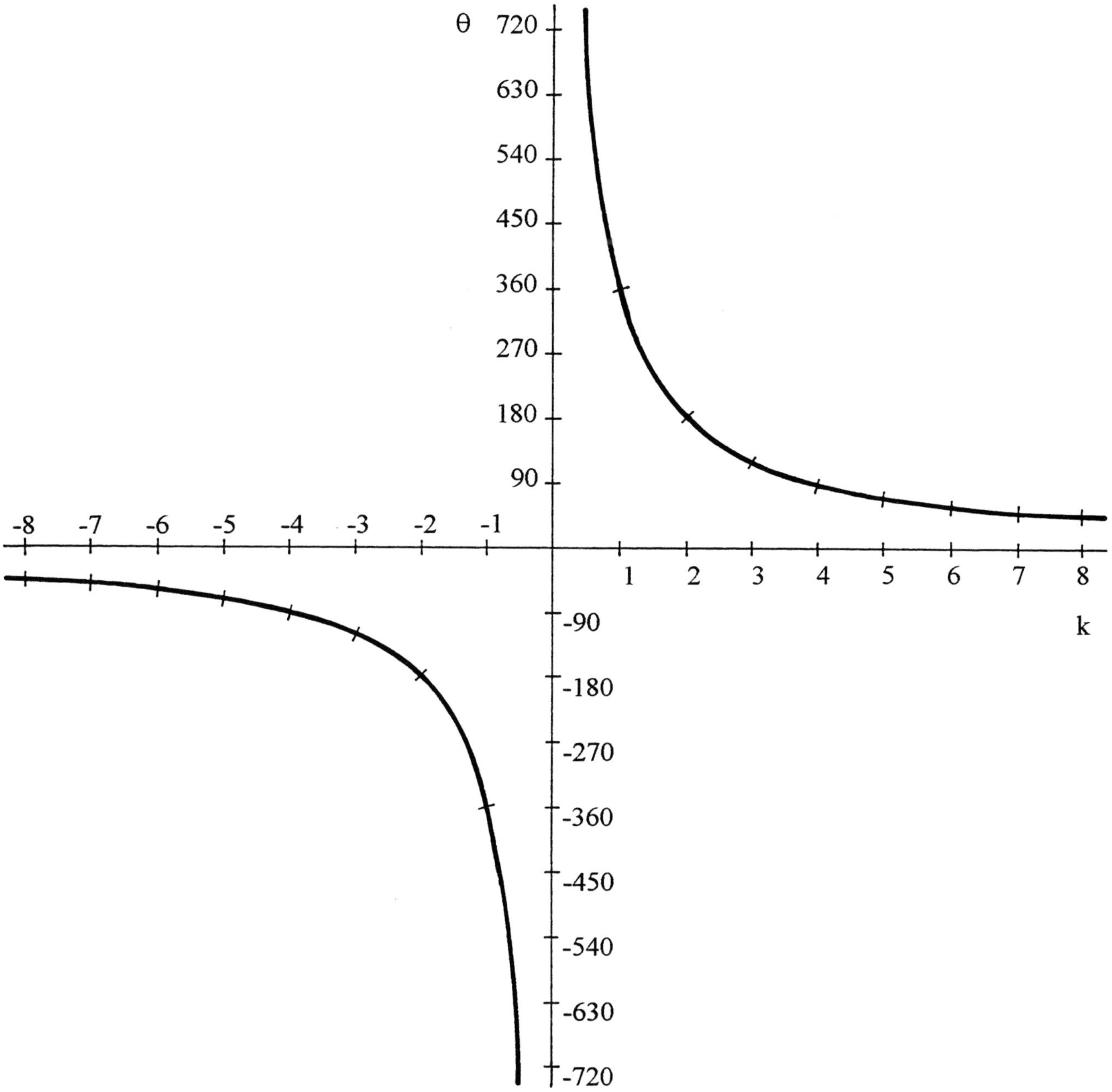

Fig.4: Graph of $\theta = 360/k$ showing Regular Polygons between $k = -8$ and $k = 8$

STAR POLYGONS

We have now dealt with both positive and negative integer values of k, which might sensibly seem as far as we can go. It is surely not possible to have fractional sided polygons?

Our graph of θ = 360/k, however, does not know this and appears to be a continuous function away from its asymptotes, as shown in Figure 4.

The results of trying values of k in halves, thirds, quarters, fifths, sixths and sevenths are given in Tables 3, 4, 5, 6, 7 and 8 respectively.

What we get are termed star polygons on account of their general shape, and it is usual to denote k by a symbol, say 5/2, rather than 2.5 or 2½. In general then k can be taken to equal n/d.

Thinking about this in terms of vectors, if, for instance, k is a whole number of halves, then it is going to take two revolutions, a total rotation of 720°, to get back to the start. If k is in fifths then five revolutions, 1800°, will be needed.

In general we can see from the tables that n represents the number of vertices of the star polygon, whilst d is the number of revolutions needed to complete it.

The star polygons can thus be seen as a more general case of polygon with regular polygons occurring as nothing more than a special case of star polygon where d = 1.

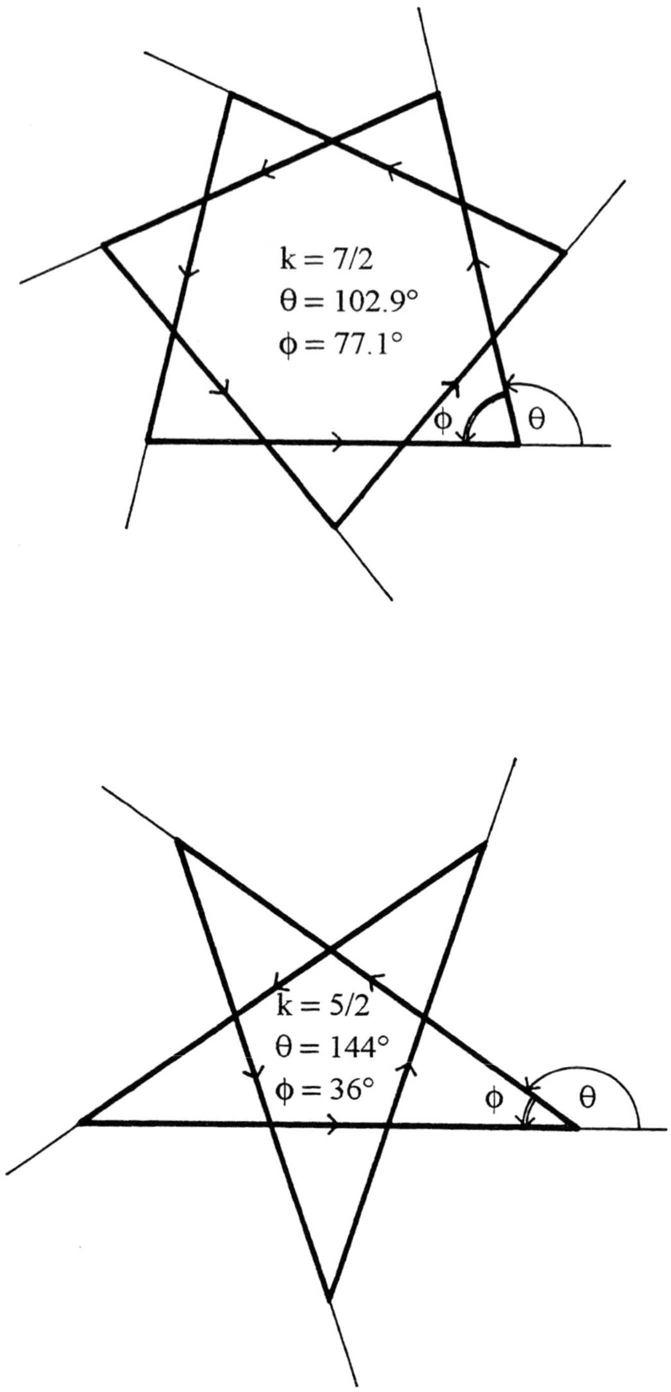

Fig.5: Generation of Star Polygons

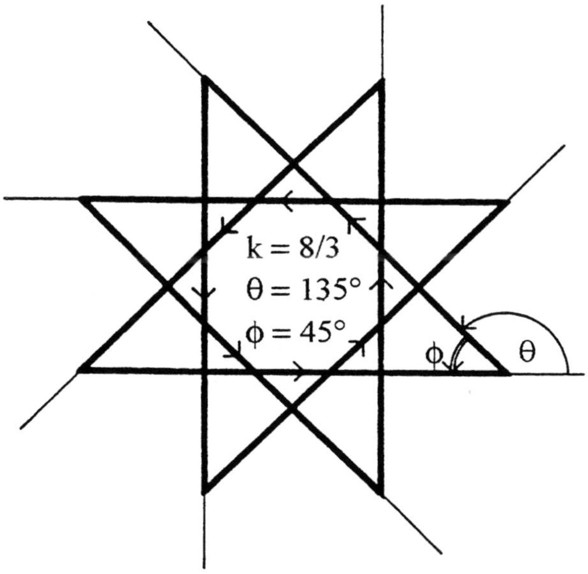
k = 8/3
θ = 135°
ϕ = 45°
ϕ
θ

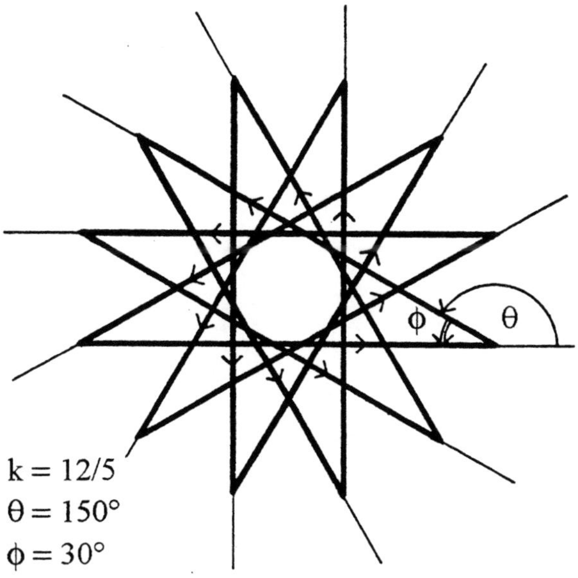
ϕ
θ
k = 12/5
θ = 150°
ϕ = 30°

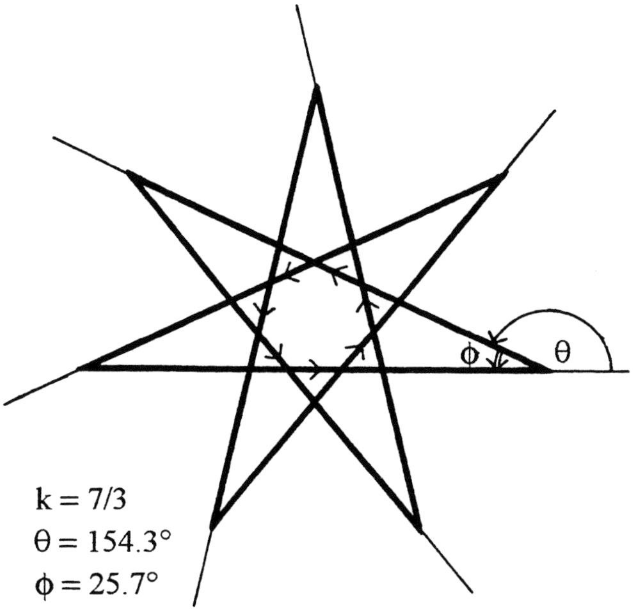
ϕ
θ
k = 7/3
θ = 154.3°
ϕ = 25.7°

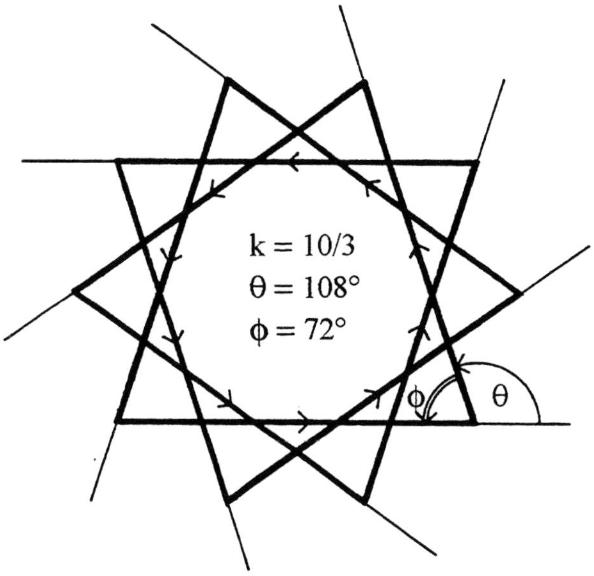
k = 10/3
θ = 108°
ϕ = 72°
ϕ
θ

	k = n/d	θ	ϕ		Polygon
	∞	0	180	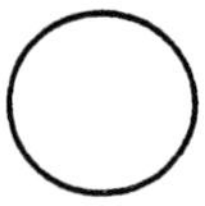	Circle
	8/2	90	90		Octagram
	7/2	102.9	77.1		Heptagram
	6/2	120	60		Hexagram
	5/2	144	36		Pentagram
	4/2	180	0		Cross
	3/2	240	−60		Triangle
	2/2	360	−180	· ·	2 Points
	1/2	720	−540	·	Point
	0	∞	∞		O

Table 3: Star Polygons of two revolutions

	k = n/d	θ	ϕ		Polygon
	∞	0	180	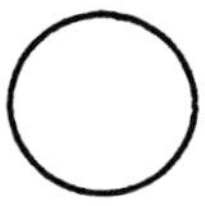	Circle
	8/3	135	45		Octagram
	7/3	154.3	25.7		Heptagram
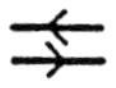	6/3	180	0		Hexacross
	5/3	216	−36		Pentagram
	4/3	270	−90		Square
	3/3	360	−180		3 Points
	2/3	540	−360		Digon
	1/3	1080	−900		Point
	0	∞	∞		O

Table 4: Star Polygons of three revolutions

	k = n/d	θ	ϕ		Polygon
	∞	0	180	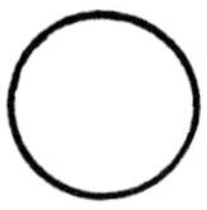	Circle
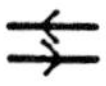	8/4	180	0		Octacross
	7/4	205.7	−25.7		Heptagram
	6/4	240	−60		Hexagram
	5/4	288	−108	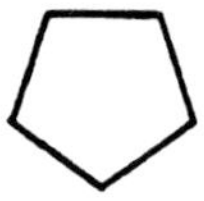	Pentagon
	4/4	360	−180		4 Points
	3/4	480	−300		Triangle
	2/4	720	−540		2 Points
	1/4	1440	−1260		Point
	0	∞	∞		O

Table 5: Star Polygons of four revolutions

	k = n/d	θ	ϕ		Polygon
	∞	0	180	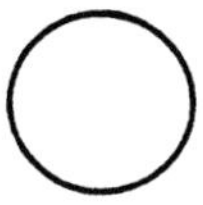	Circle
	8/5	225	−45		Octagram
	7/5	257.1	−77.1		Heptagram
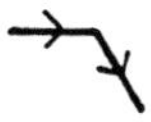	6/5	300	−120		Hexagon
	5/5	360	−180	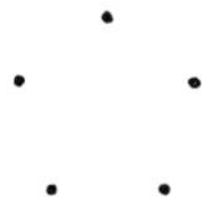	5 Points
	4/5	450	−270		Square
	3/5	600	−420	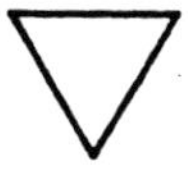	Triangle
	2/5	900	−720		Digon
	1/5	1800	−1620		Point
	0	∞	∞		O

Table 6: Star Polygons of five revolutions

	$k = n/d$	θ	ϕ		Polygon
	∞	0	180	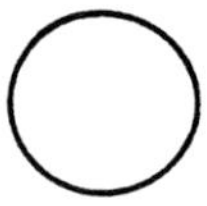	Circle
	8/6	270	−90		Octagram
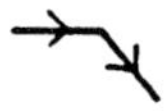	7/6	308.6	−128.6		Heptagon
	6/6	360	−180		6 Points
	5/6	432	−252	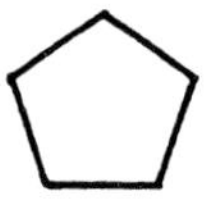	Pentagon
	4/6	540	−360		Cross
	3/6	720	−540		3 Points
	2/6	1080	−900		2 Points
	1/6	2160	−1980		Point
	0	∞	∞		O

Table 7: Star Polygons of six revolutions

	k = n/d	θ	ϕ		Polygon
	∞	0	180	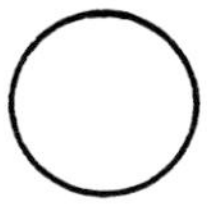	Circle
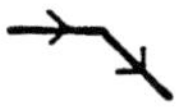	8/7	315	−135	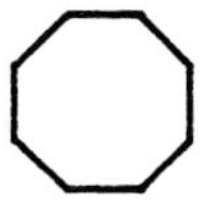	Octagon
	7/7	360	−180	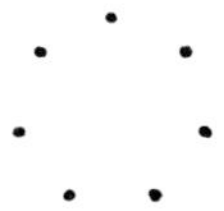	7 Points
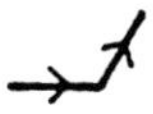	6/7	420	−240	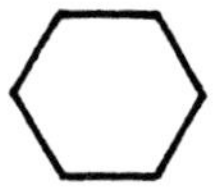	Hexagon
	5/7	504	−324		Pentagram
	4/7	630	−450		Square
	3/7	840	−660		Triangle
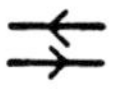	2/7	1260	−1080		Digon
	1/7	2520	−2340		Point
	0	∞	∞		O

Table 8: Star Polygons of seven revolutions

COMPOUND POLYGONS

In a number of cases of star polygon in the preceding tables we have duplicated a value of k that was originally used for a regular or star polygon, but which now appears in the form cx/cy. This produces a compound polygon, which in general consists of c polygons, x/y, arranged around a total of cx vertices.

Figure 6 illustrates some examples of these and it should be noted that if c itself is not a prime number, and therefore has factors, the compound can be regarded as a compound of a compound, e.g. in the case of 12/4, this can either be seen as four 3/1s or alternatively two 6/2s.

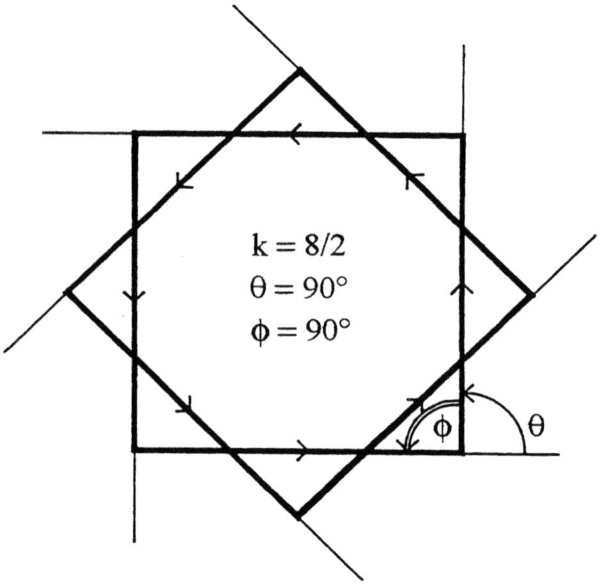

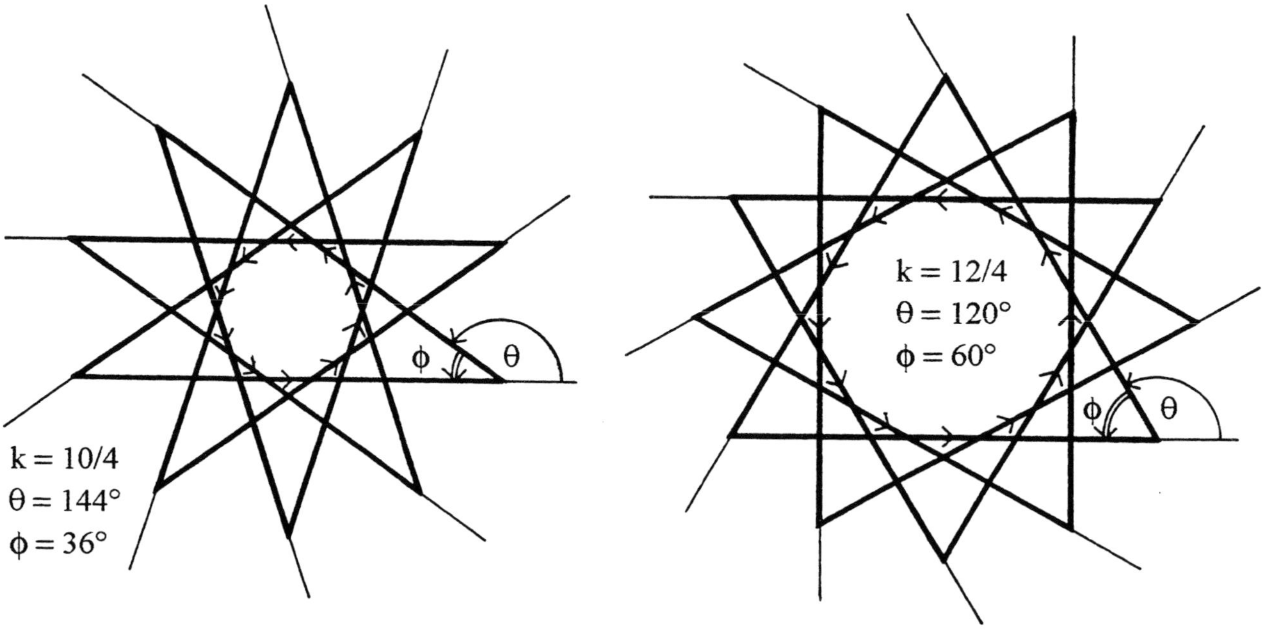

Fig.6: Generation of Compound Polygons

CROSS POLYGONS

A special case of compound polygon arises when k = 2, the components of the compound being digons. The result is a cross polygon consisting of c digons arranged diametrically.

Figure 7 illustrates some examples of these ranging from the simplest cross polygon 4/2 to the more complex 8/4 which again can either be regarded as four 2/1s or as a compound of a compound in the form of two 4/2s.

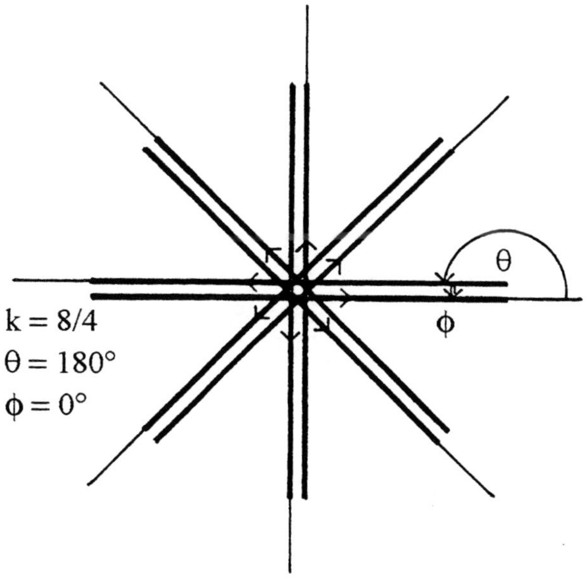

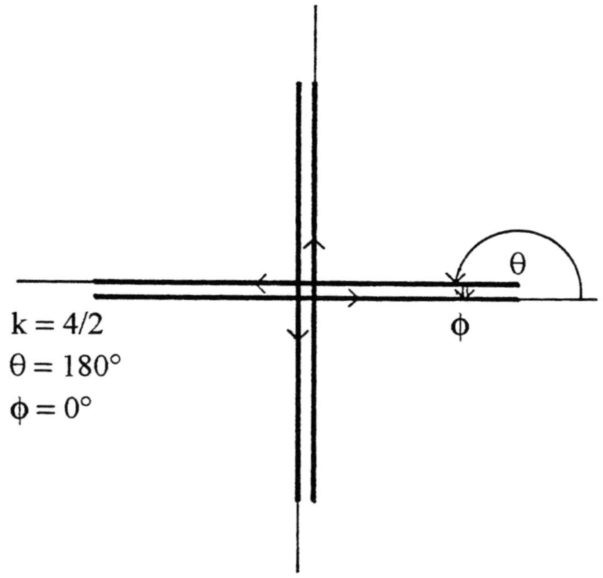

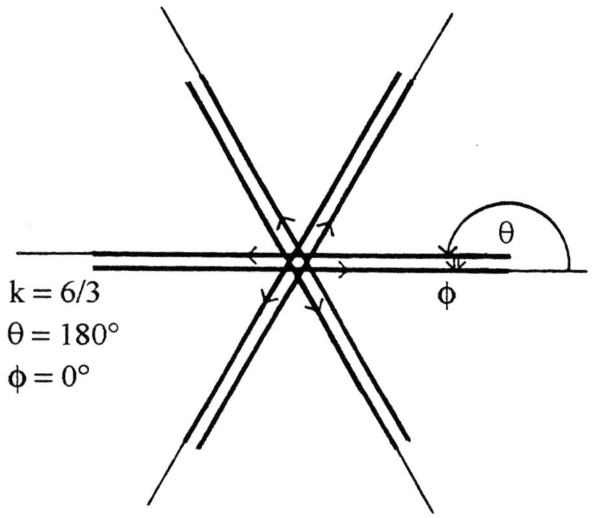

Fig.7: Generation of Cross Polygons

INVERSE POLYGONS

Also apparent in the preceding tables will be a number of seemingly negative polygons which have a clockwise direction to their vectors, but with a value of k between 1 and 2, rather than a negative value.

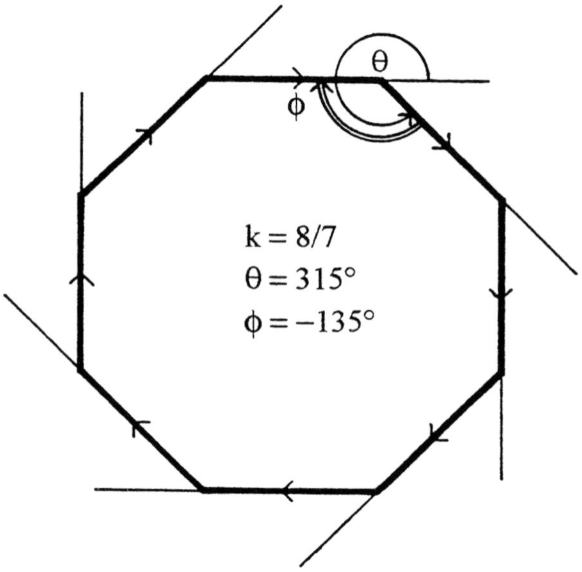

These are the inverse polygons which differ subtly from the negative polygons as we shall see later when we come to truncation.

In general any positive polygon k = n/d will have an inverse polygon k' = n'/d', where n'= n and d' = n–d, i.e. k' = n/(n–d). This results in the property that 1/k + 1/k' = 1.

A consequence of this for cross polygons, with n = 2d, is that they are self-inverses.

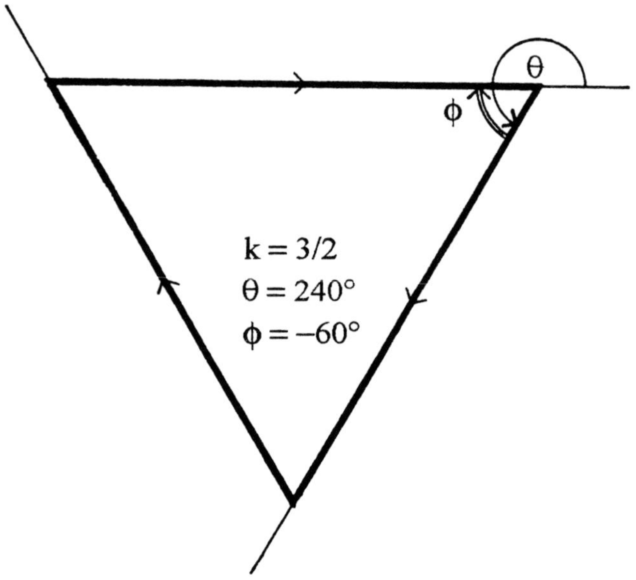

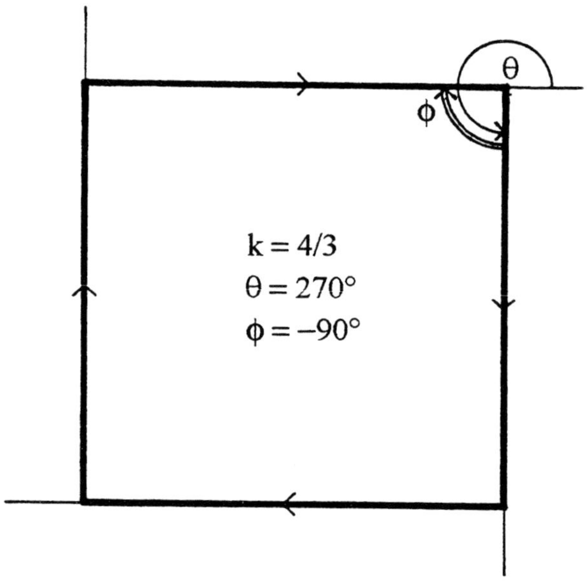

Fig.8: Generation of Inverse Polygons

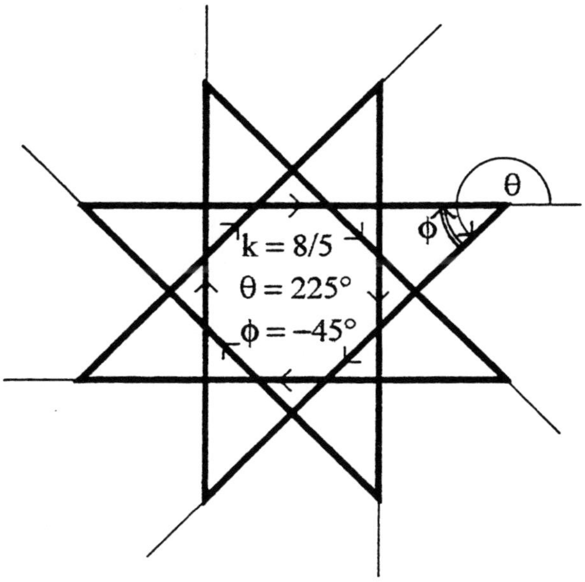
θ
φ
k = 8/5
θ = 225°
φ = −45°

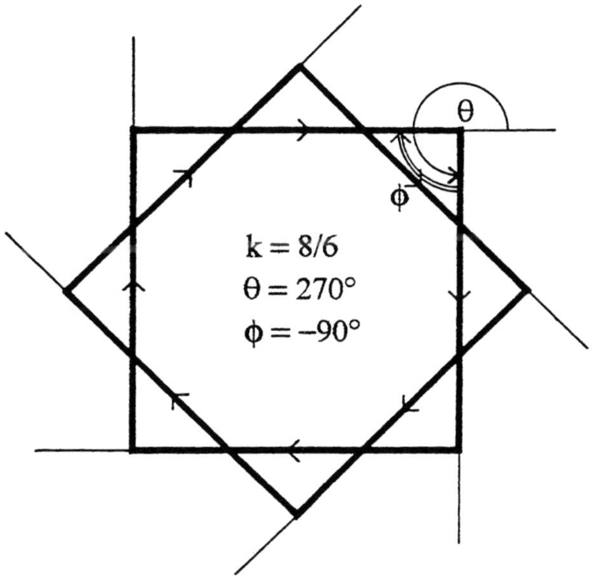
θ
φ
k = 8/6
θ = 270°
φ = −90°

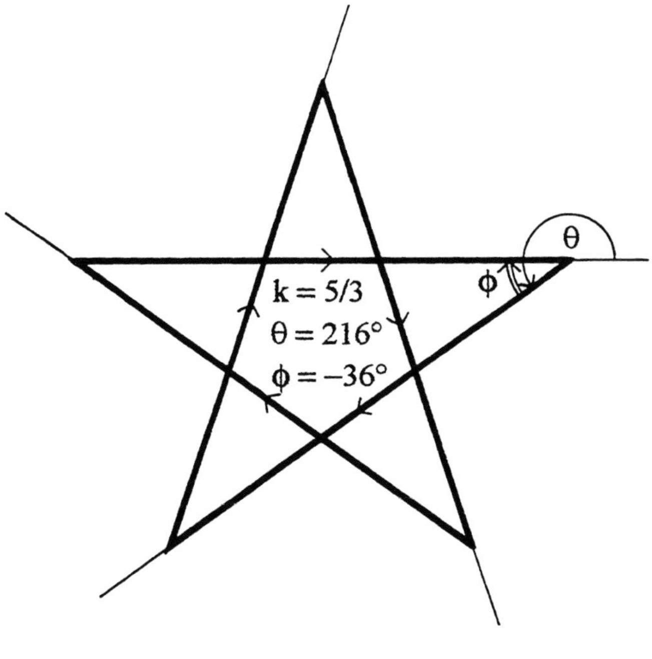
θ
φ
k = 5/3
θ = 216°
φ = −36°

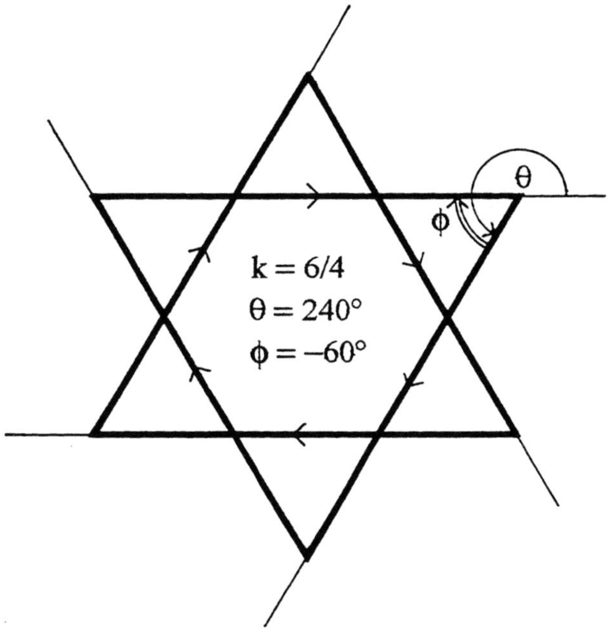
θ
φ
k = 6/4
θ = 240°
φ = −60°

POLYGONS GENERALLY

We have now covered all possible values of k for which n and d remain as small integers. It is useful to summarise what we have found by plotting n against d, as in Table 9.

What we find is that the top left part of the diagram down to a diagonal line where k = 2 is occupied by star polygons with the special case of regular polygons with d = 1 running off to infinity up the left hand side.

Along the line k = 2 we find the cross polygons, with to their right the inverse polygons, mirroring exactly those on the left down as far as the line where k = 1.

Along this line k = 1, we have k = n/n, with each of the n points a polygon should start at getting no further than itself.

Below k = 1 the entire process repeats itself again and again, firstly between k = 1 and k = ½, with a central line of cross polygons at k = ⅔, and once more between k = ½ and k = ⅓ with another line of cross polygons at k = ⅖ and so on.

This repetition is the result of cycles of 360° in the value of θ and can be seen more clearly by plotting our graph of θ = 360/k instead as one of θ = 360(1/k) as shown in Figure 9. Here each time θ goes beyond 360° we can subtract 360° and end up with a sawtooth graph with a series of equal sections, each of which represents one complete? set of polygons and inverses.

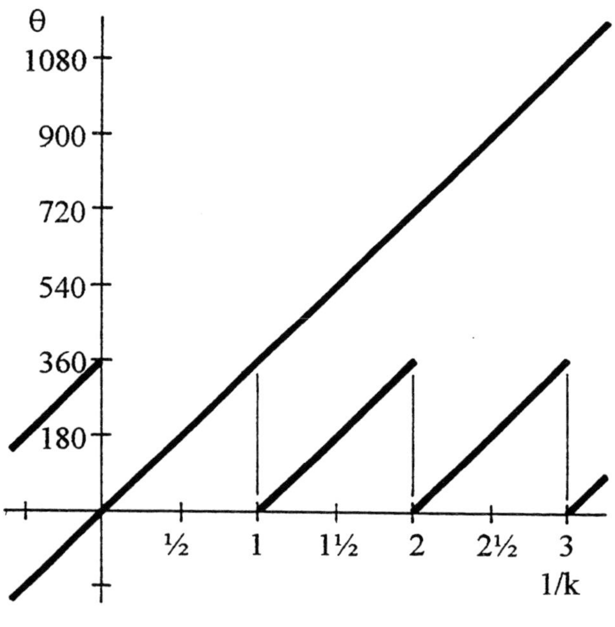

Fig.9: Graph of θ = 360(1/k)

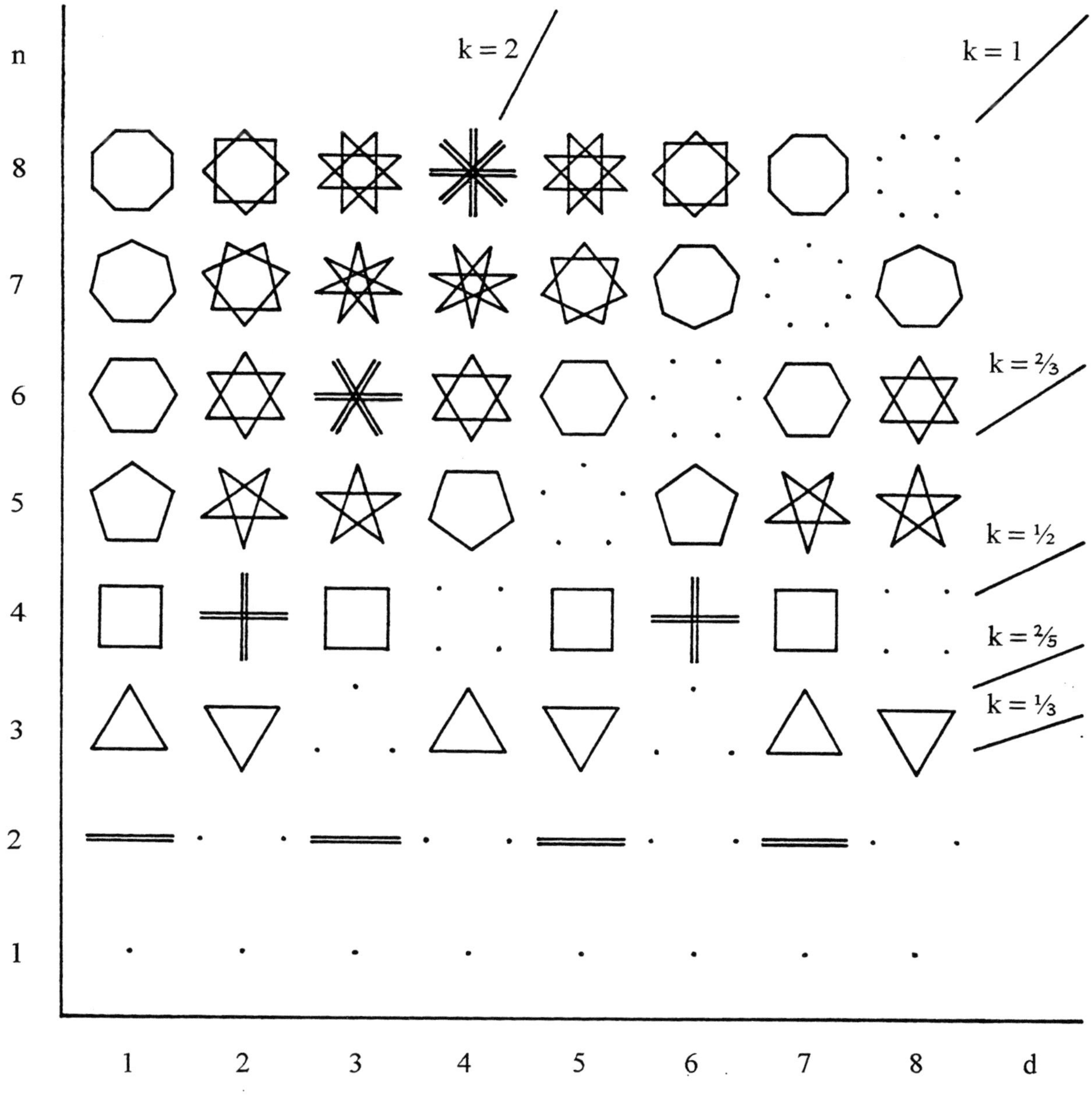

Table 9: The Polygons k = n/d

THE COMPLETE? SET

Because of this repetition every 360° we can regard any polygon n/d as generally equivalent to n/(d±cn) and simplify matters by reducing Table 9 to only include the set of polygons between k = ∞ and k = 1, as shown in Table 10.

For any value of n the different values of d can be interpreted by placing n points equidistantly around a circle as shown in Figure 10. The vector needed to generate any particular polygon will be found by proceeding from our starting point to the dth point anticlockwise around the circle, for which $\theta_d = (d+1)\,\theta_1/2$.

We can clearly see in the symmetry of the diagrams how an inverse pair share the same vector but in opposite directions around the circle, and that if n is even the diagonal, d = n/2, will yield the cross polygon.
If d > n then we pass around the circle more than once to reach the dth point and thereby repeat the cycle as already observed.

At the point where d = n, each point around the circle yields just a point polygon there, this situation also applying in the curious case where d = 0, where theoretically k = ∞, and we have an equivalent circle.

If we give d a negative value and thus go backwards around the circle, we can equate the negative polygons, previously considered, with the inverse polygons.
By going further round the circle past the diagonal we get the negative inverse polygons which are effectively positive!

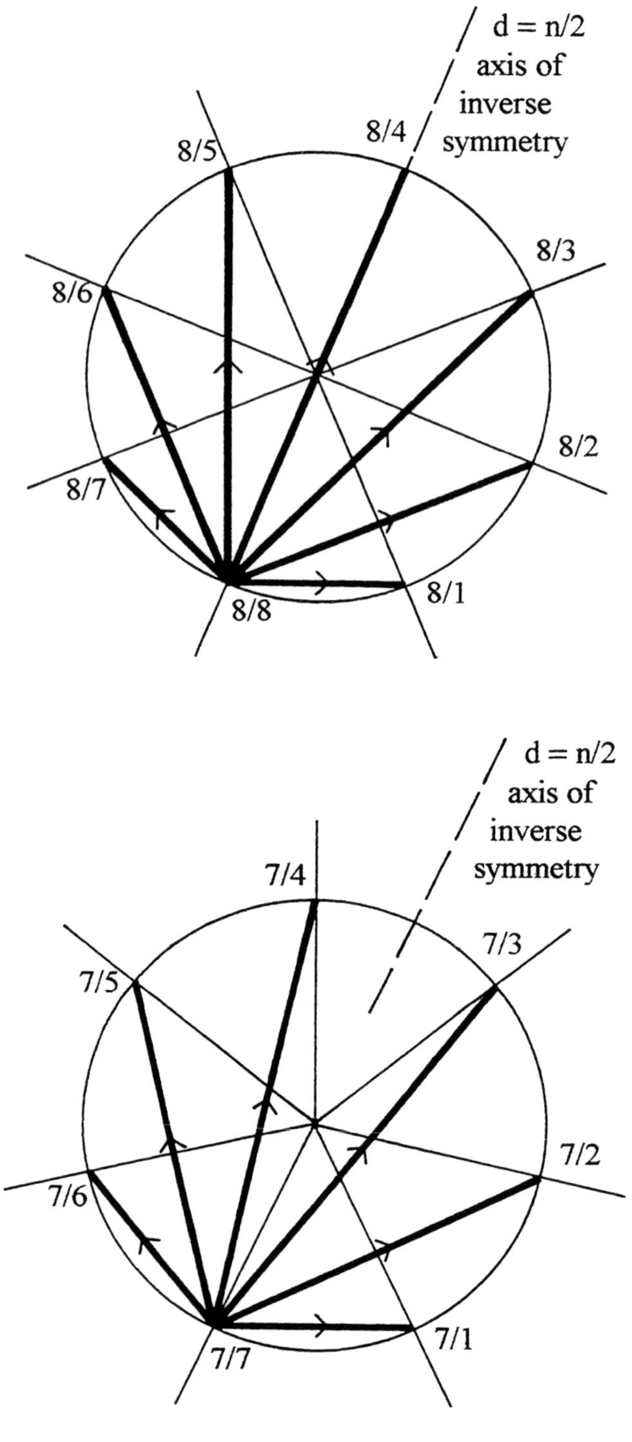

Fig. 10: Interpretation around a Circle

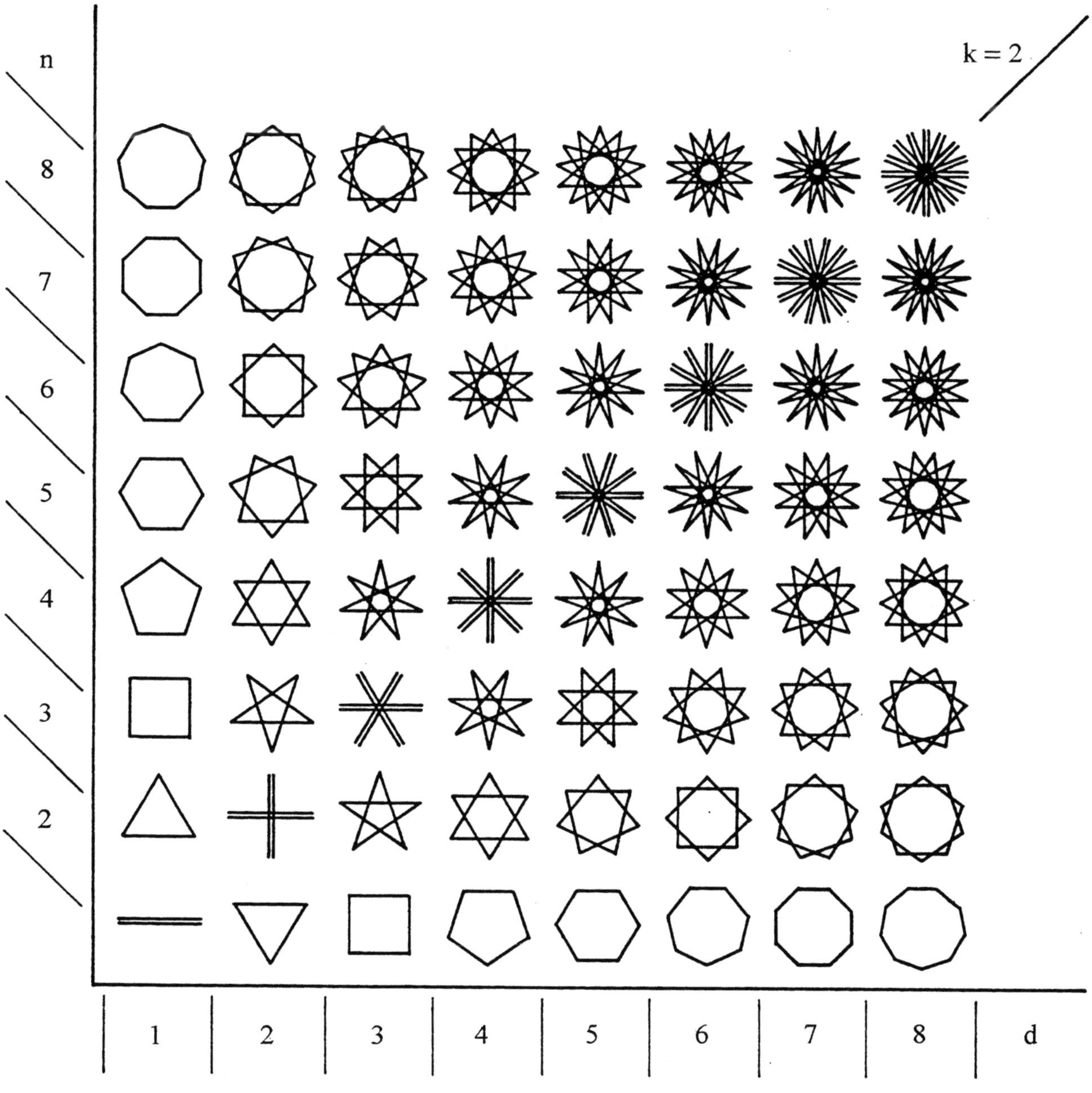

Table 10: Part of the Complete? Set of Polygons

INSCRIBING POLYGONS

From Figure 10 we can see that within each regular polygon, k = n/1, we can fill in sets of diagonals in turn towards the centre. These inscriptions produce a series of star polygons, k = n/2, k = n/3 etc. as shown in Figures 11 and 12. The polygon k = n/d is thus the (d − 1)th inscription of the regular polygon k = n/1.

Within the centre of each layer in turn the original polygon is recreated on a smaller scale and rotated through 360/2n°, i.e. θ/2.

In the case of the (n/2 − 1)th inscription of an even valued regular polygon we get the cross polygon, k = n/(n/2). The central figure in this instance is infinitely small, as the paired diagonals effectively coexist.

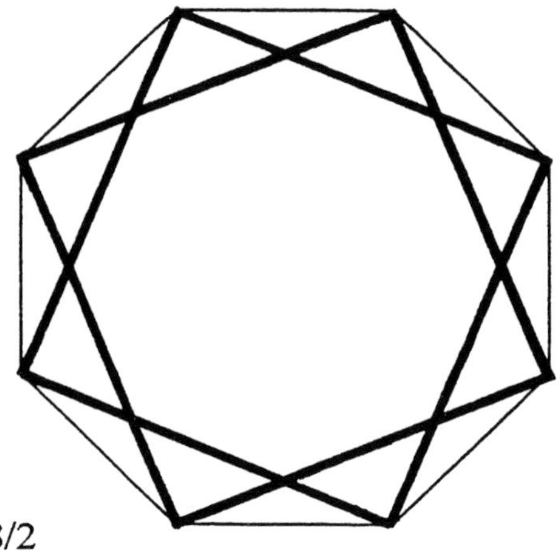

1st
k = 8/2

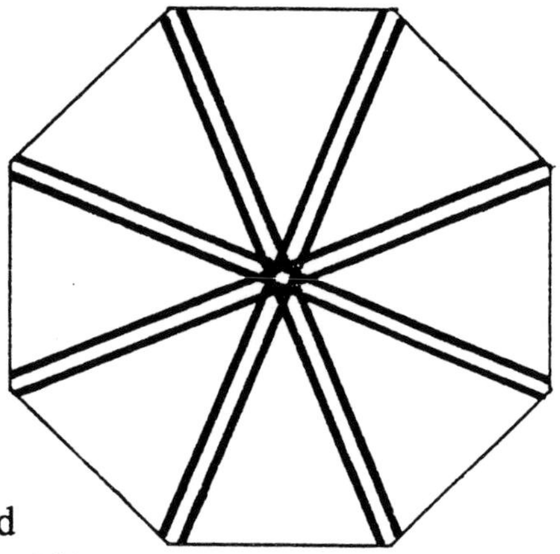

3rd
k = 8/4

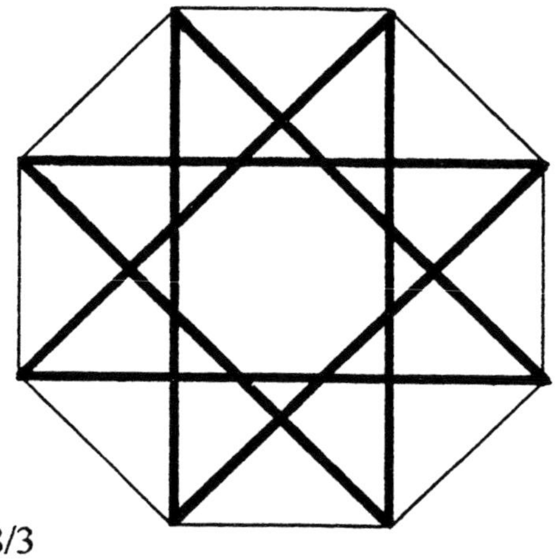

2nd
k = 8/3

Fig.11: Inscriptions of Octagon

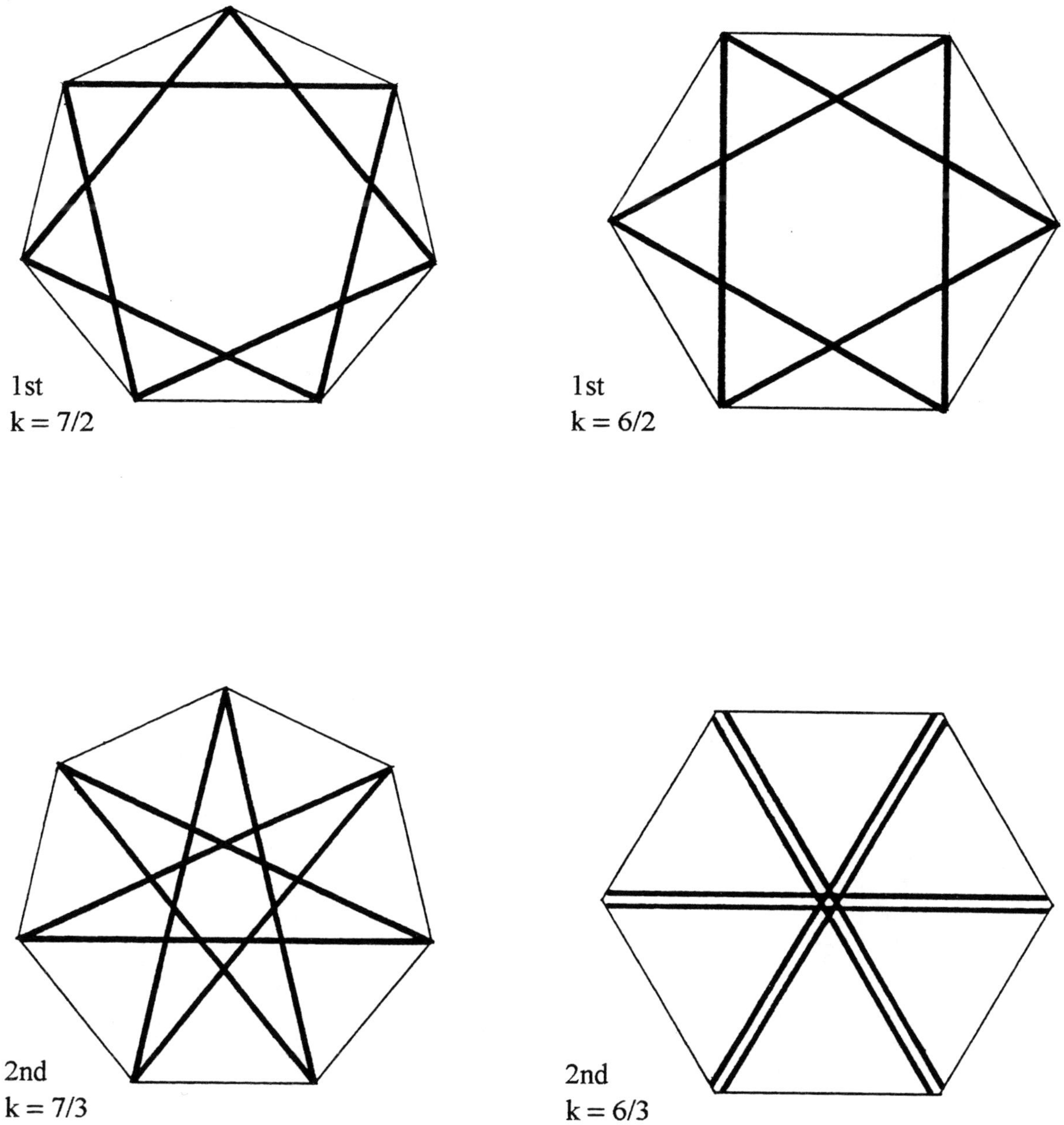

Fig. 12: Inscriptions of Heptagon and Hexagon

STELLATING POLYGONS

By starting with the central polygon the inscription of polygons can be reversed by a process called stellation, whereby the existing edges of the regular polygon, k = n/1, are extended until they cross, again producing a series of star polygons k = n/2, k = n/3 etc. as shown in Figures 13 and 14. The polygon k = n/d is thus also the (d – 1)th stellation of the regular polygon k = n/1.

Again when n is an even number the (n/2 – 1)th stellation will be the cross polygon, k = n/(n/2), now infinitely large to enable the finitely spaced parallel diagonals to eventually cross.

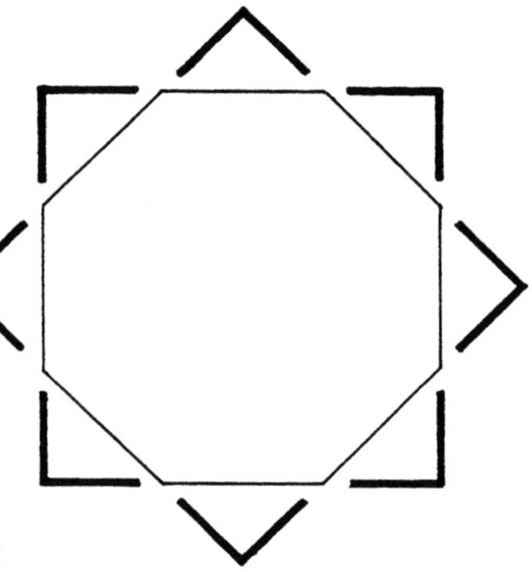

1st
k = 8/2

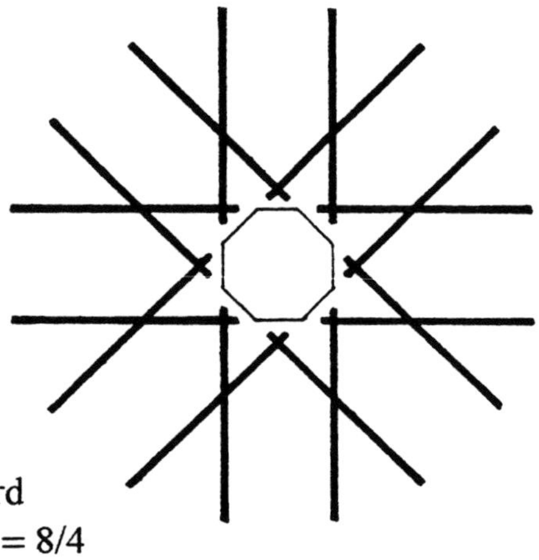

3rd
k = 8/4

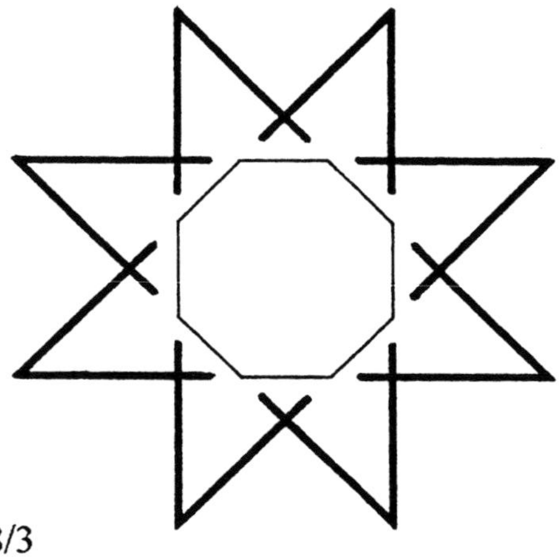

2nd
k = 8/3

Fig. 13: Stellations of Octagon

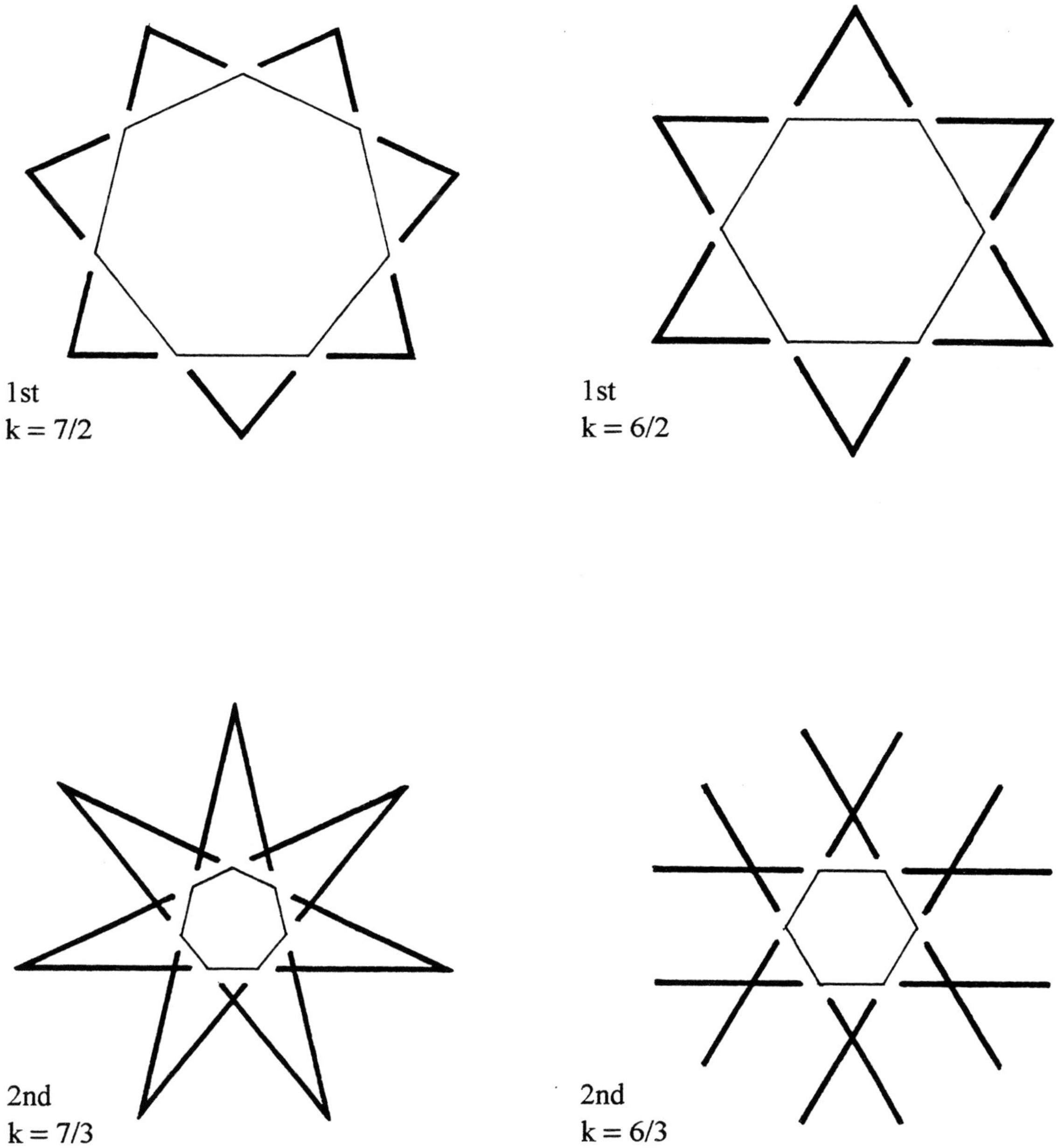

Fig. 14: Stellations of Heptagon and Hexagon

DENSITY OF POLYGONS

We have seen from the above that any star polygon, k = n/d, can generally be regarded as the (d – 1)th inscription or stellation of a regular polygon, k = n/1.

The quantity d, which up to now has merely been a denominator, can now be seen to represent the 'density' of the polygon.

This equates to the number of revolutions needed to create the polygon and is effectively the number of edges one has to cross to get from the outside to the centre of the polygon. During the process of building up a star polygon by stellation this quantity increases by one for each additional set of extended edge crossings added.

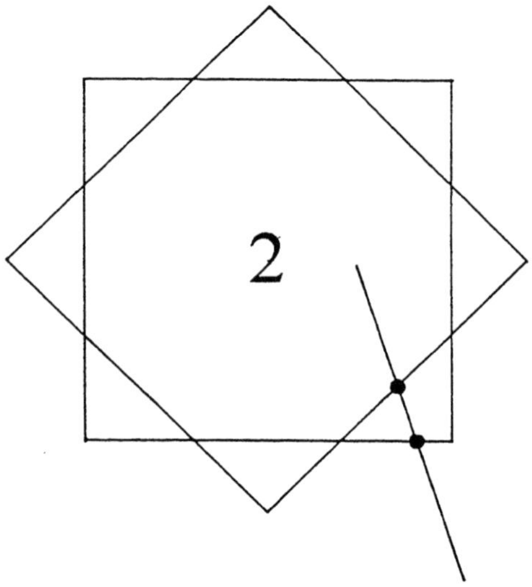

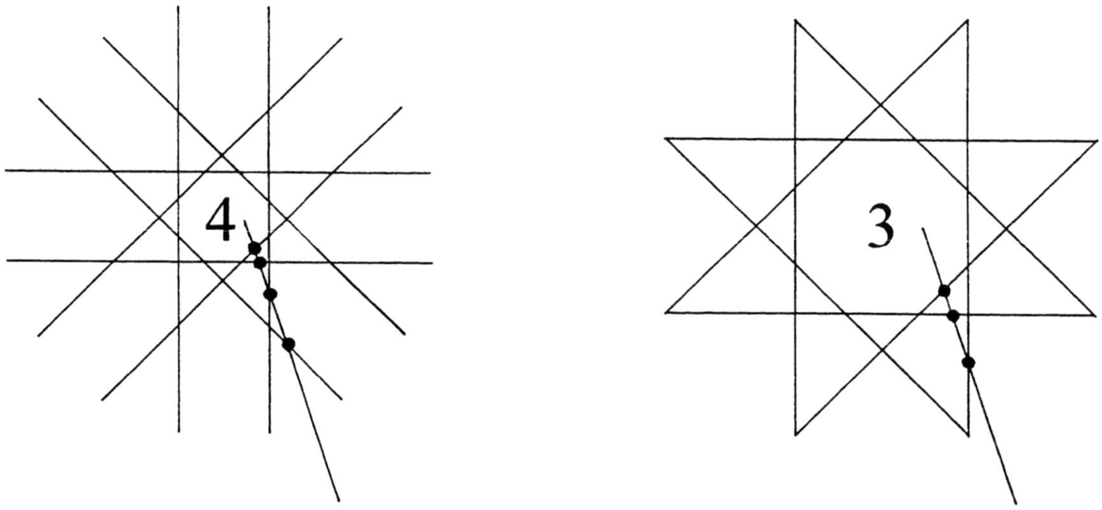

Fig.15: Density of Polygons

DENSITY OF INVERSE POLYGONS

When d > n/2, k = n/d cannot be regarded as an inscription or stellation of a regular polygon.

The quantity d, however, still equates to the number of revolutions needed to create the polygon. The angle between consecutive vectors is now turned through externally and can be shown as in Figure 16.

Here we can see that d is still effectively the number of edges one has to cross to get from the outside to the centre of the polygon.

N.B. in cases where n is odd, two different numbers of edge crossings can be found, their mean equating to d.

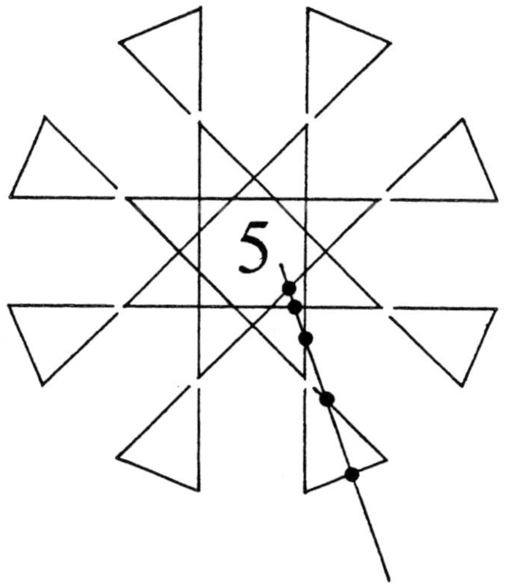

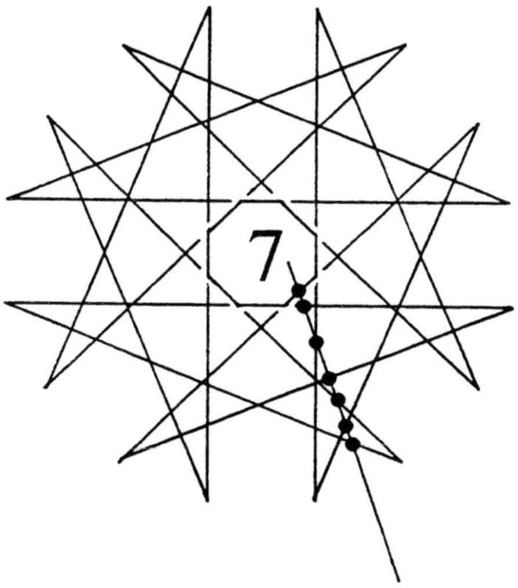

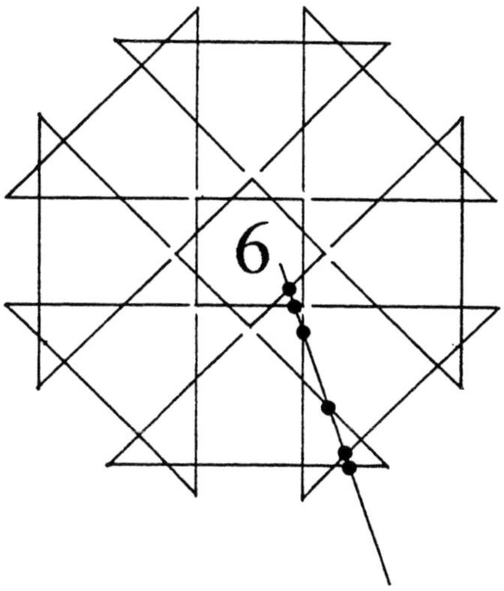

Fig.16: Density of Inverse Polygons

DUAL POLYGONS & AREAS

In an operation similar to inscription we can join up the centres of the edges of a polygon and produce its dual, which is always another polygon of the same type, as shown in Figure 17.

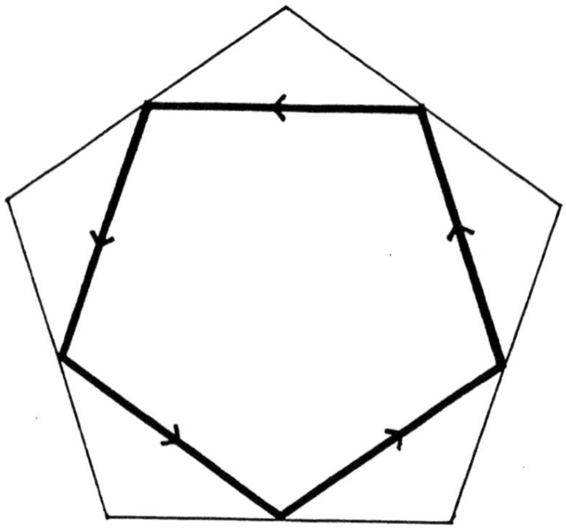

Fig.17: Generation of Dual Polygons

It is normal to ascribe the length 2 units to the edges of a polygon and the length of the edges of the dual at this scale will be of use later when constructing 'vertex figures' for the polyhedra. Using our vector notation this dual edge length, m, can be seen to be half the length of the vector that sums two consecutive edges of the polygon, i.e. $m = 2\cos(\theta/2)$. See Figure 18. This is summarised for the polygons more commonly found in polyhedra in Table 11.

Also shown in Table 11 are the areas of the same polygons of edge length 2 units, given by the formula:
$A = na = n\tan(\phi/2) = n\tan(90(n-2d)/n)$.

This is fairly straightforward for regular polygons where a series of triangular areas are simply summed. Their inverses are also simple with areas of equal magnitude but negative sign, a property suggestive of area having a directional magnitude perpendicular to the plane, such that any polygon viewed as anticlockwise (e.g. an inverse polygon from behind) will have positive area.

The areas given for star polygons are not quite so simple as the constituent triangles overlap in the centre giving a result greater than the immediately apparent area. This can be interpreted in terms of the increasing density towards the centre of a star polygon being representative of additional layers.

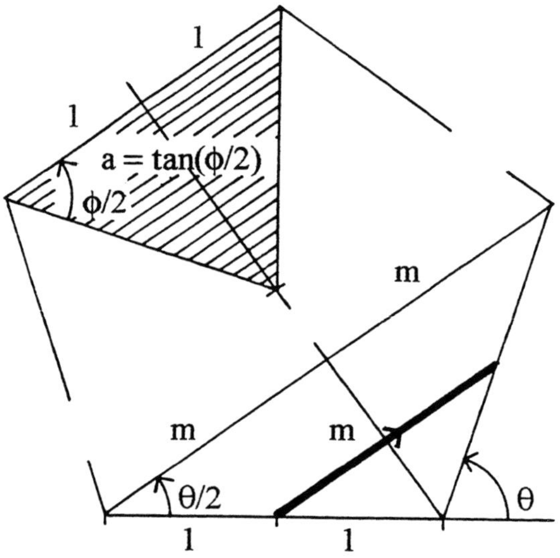

Fig.18: Constructions for Dual Edge Length and Area

k = n/d	θ/2	2cosθ/2	= m	ϕ/2	tanϕ/2 = a	n(a) = A
∞	0	2.000	√(2+2)	90	∞	∞
12/1	15	1.932	√(2+√3)	75	3.732	12(√(2+√3)/√(2−√3))
10/1	18	1.902	√(2+τ)	72	3.078	10(√(2+τ)/√(2−τ))
8/1	22.5	1.848	√(2+√2)	67.5	2.414	8(√(2+√2)/√(2−√2))
6/1	30	1.732	√(2+1)	60	1.732	6(√(2+1)/√(2−1))
5/1	36	1.618	√(2+(τ−1))	54	1.376	5(√(2+(τ−1))/√(2−(τ−1)))
4/1	45	1.414	√(2+0)	45	1.000	4(√(2+0)/√(2−0))
10/3	54	1.176	√(2−(τ−1))	36	0.727	10(√(2−(τ−1))/√(2+(τ−1)))
3/1	60	1.000	√(2−1)	30	0.577	3(√(2−1)/√(2+1))
8/3	67.5	0.765	√(2−√2)	22.5	0.414	8(√(2−√2)/√(2+√2))
5/2	72	0.618	√(2−τ)	18	0.325	5(√(2−τ)/√(2+τ))
12/5	75	0.518	√(2−√3)	15	0.268	12(√(2−√3)/√(2+√3))
4/2	90	0.000	√(2−2)	0	0.000	4(√(2−2)/√(2+2))
5/3	108	−0.618	−√(2−τ)	−18	−0.325	−5(√(2−τ)/√(2+τ))
3/2	120	−1.000	−√(2−1)	−30	−0.577	−3(√(2−1)/√(2+1))
4/3	135	−1.414	−√(2−0)	−45	−1.000	−4(√(2+0)/√(2−0))
5/4	144	−1.618	−√(2+(τ−1))	−54	−1.376	−5(√(2+(τ−1))/√(2−(τ−1)))
6/5	150	−1.732	−√(2+1)	−60	−1.732	−6(√(2+1)/√(2−1))

Table 11: Dual Edge Lengths and Areas of Common Polygons (τ = (1+√5)/2, the golden section)

TRUNCATING POLYGONS

When considering later on the relationships between various three dimensional polyhedra, the question of truncation will loom large. To understand this fully we will need an appreciation of the polygons and their behaviour under truncation.

In the simplest terms the truncation of a regular polygon involves the cutting off of its vertices to produce new edges. This process can be continued until eventually the centre-point of the original polygon is all that remains. The point at which these new edges first meet each other, extinguishing the original edges at their mid-points, produces the dual of the polygon with the same number of edges as the original.

In the course of the process leading up to the dual, we reach a partway stage at which another regular polygon is encountered, generally one with twice as many edges, these alternating around the circumference as new edges and remains of the old edges.

Returning to our original vector model we can interpret truncation as the production of a truncated polygon by the halving of the external angle θ, two such angles now being required to get around from one old edge to the next, with a new edge inbetween.

The polygon k, θ, ϕ thus truncates to produce a new polygon k*, θ*, ϕ* where $k^* = 2k$, $\theta^* = \theta/2$, $\phi^* = 180 - \theta/2$.
See Table 12.

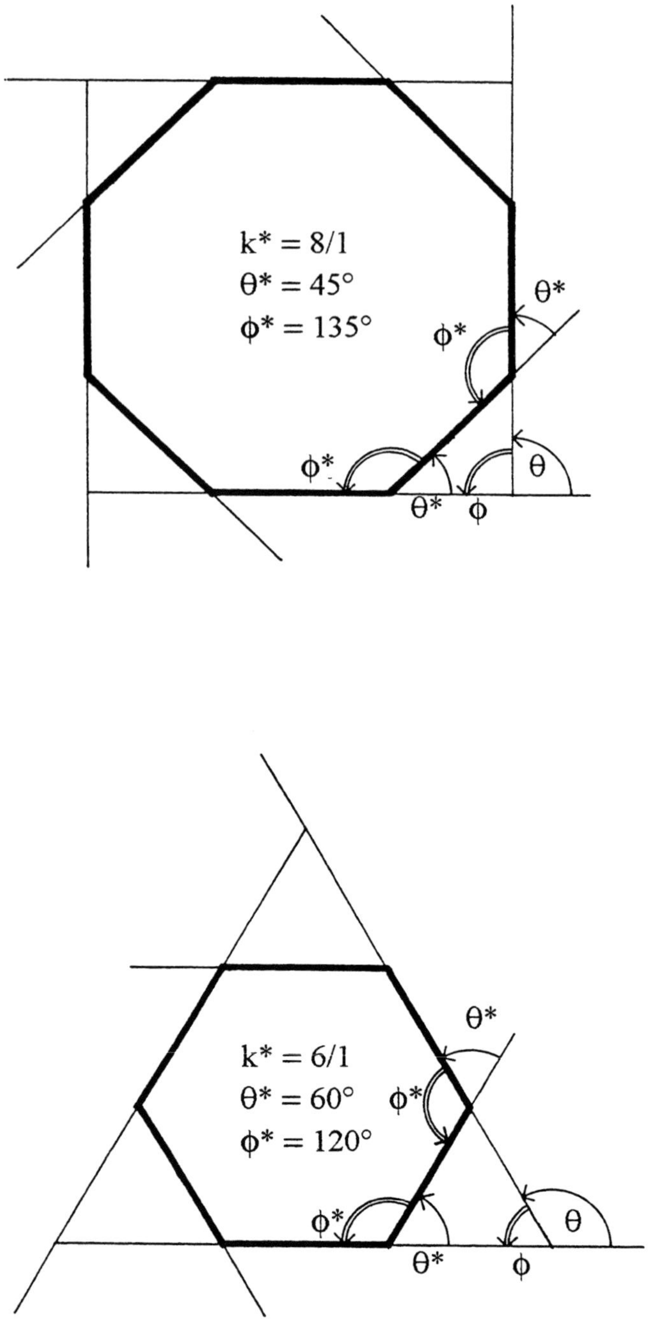

Fig.19: Truncation of Polygons

k	k*	θ*	φ*		Truncation
∞	∞	0	180	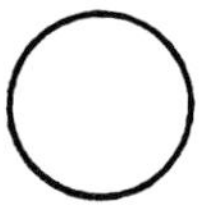	Circle
8/1	16/1	22.5	157.5	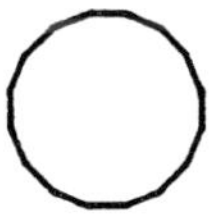	Hexadecagon
7/1	14/1	25.7	154.3	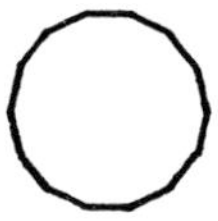	Tetradecagon
6/1	12/1	30	150	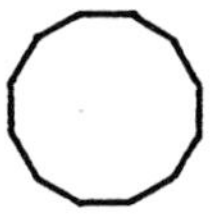	Dodecagon
5/1	10/1	36	144	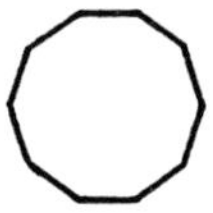	Decagon
4/1	8/1	45	135	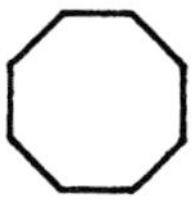	Octagon
3/1	6/1	60	120	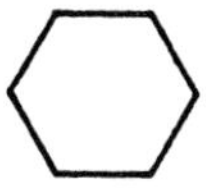	Hexagon
2/1	4/1	90	90		Square
1/1	2/1	180	0		Digon
0	0	∞	∞		O

Table 12: Truncated Regular Polygons

DOUBLE POLYGONS

If we attempt to truncate a negative regular polygon in the same way as above we obtain nothing more than a negative truncated polygon, a mirror image of the positive.

However if we try to truncate the star and inverse polygons using the same formulae, the results are much more interesting as shown in Table 13. Here the star polygons of two revolutions give rise to double polygons as their truncations. These all have the correct numbers of sides, i.e. twice that of the originals, but can be seen to double up so that each edge consists of one new and one old edge along the same line. These will generally be attributed the symbol ${}^2(n/1)$ to distinguish them from their parents n/2 or the regular polygons n/1.

Because of this doubling up of edges these double polygons retain the double density of their parents and can be seen in the plane as two coincident polygons. This double density is a property analogous to the half spin fermions of particle physics, which only half turn around under a rotation of 360°! Continuation of the truncation process past these figures will of course still lead on to the duals of the originals, which also retain the double density.

Further star polygons are truncated in Tables 14, 15, 16, 17 and 18, where it can be seen that generally the polygon k = n/d truncates to k* = 2n/d when d is odd and a double polygon k* = ${}^2(n/(d/2))$ when d is even. This latter produces doubles of both star and inverse polygons in their turn.

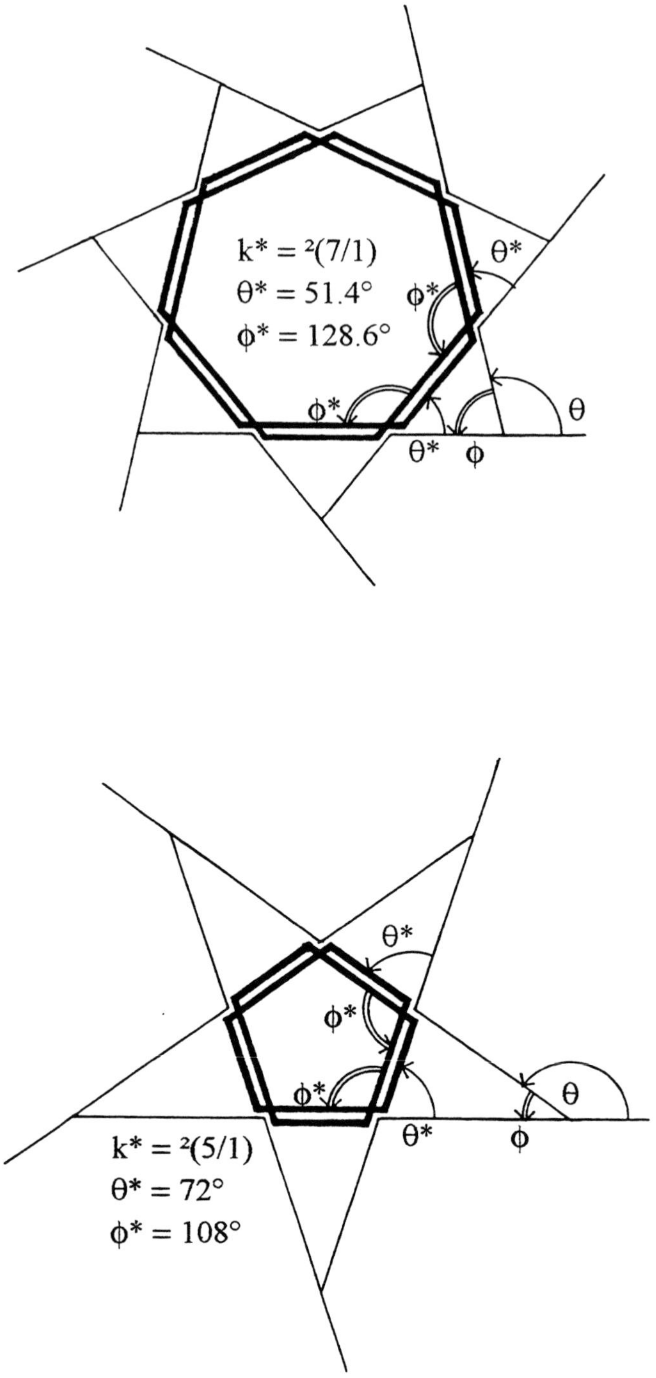

Fig.20: Truncation of Star Polygons

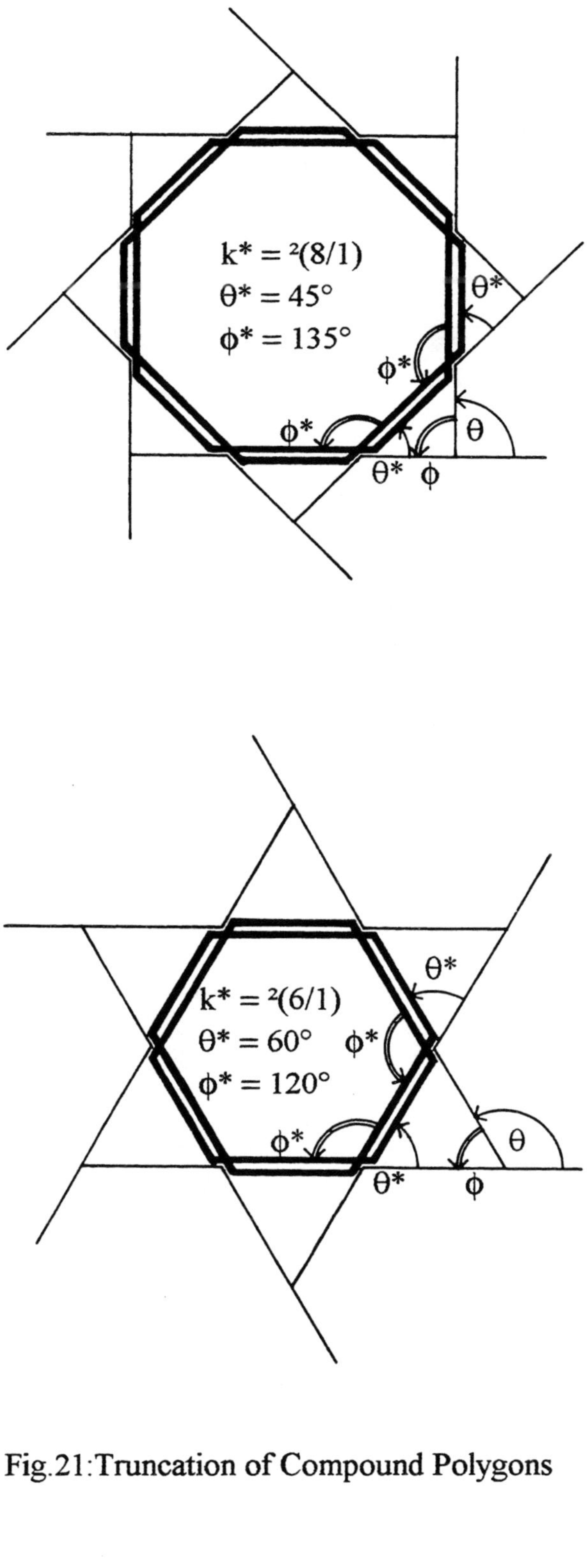

Fig.21:Truncation of Compound Polygons

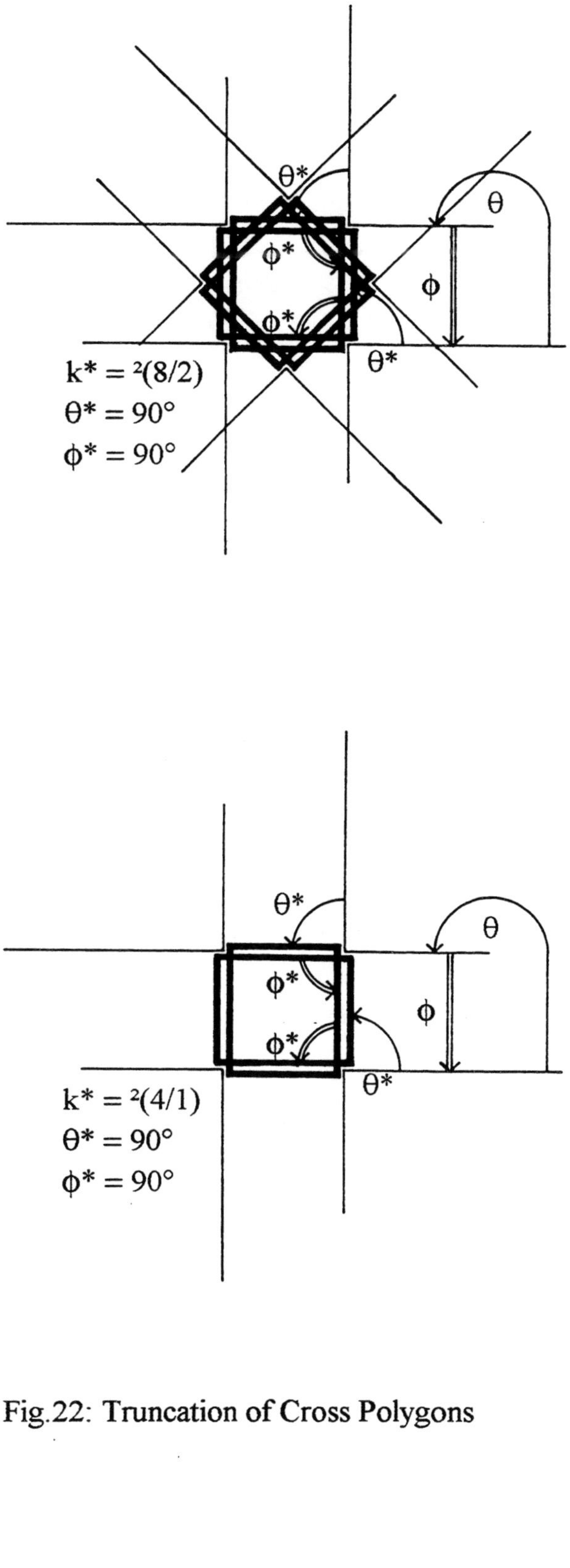

Fig.22: Truncation of Cross Polygons

k	k*	θ*	φ*		Truncation
∞	∞	0	180		Circle
8/2	$^{2}(8/1)$	45	135	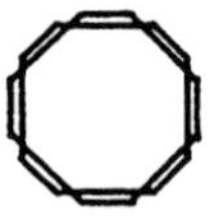	Octagon
7/2	$^{2}(7/1)$	51.4	128.6		Heptagon
6/2	$^{2}(6/1)$	60	120		Hexagon
5/2	$^{2}(5/1)$	72	108		Pentagon
4/2	$^{2}(4/1)$	90	90		Square
3/2	$^{2}(3/1)$	120	60		Triangle
2/2	$^{2}(2/1)$	180	0		Digon
1/2	$^{2}(1/1)$	360	−180	:	Point
0	0	∞	∞		O

Table 13: Truncated Star Polygons of two revolutions

k	k*	θ*	φ*		Truncation
∞	∞	0	180	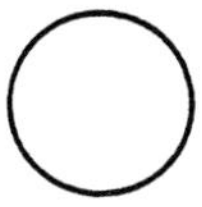	Circle
8/3	16/3	67.5	112.5		Hexadecagram
7/3	14/3	77.1	102.9		Tetradecagram
6/3	12/3	90	90		Dodecagram
5/3	10/3	108	72		Decagram
4/3	8/3	135	45		Octagram
3/3	6/3	180	0		Hexacross
2/3	4/3	270	–90		Square
1/3	2/3	540	–360		Digon
0	0	∞	∞		O

Table 14: Truncated Star Polygons of three revolutions

k	k*	θ*	ϕ*		Truncation
∞	∞	0	180		Circle
8/4	$^2(8/2)$	90	90	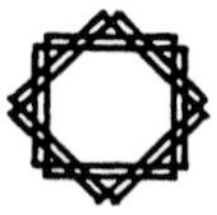	Octagram
7/4	$^2(7/2)$	102.9	77.1		Heptagram
6/4	$^2(6/2)$	120	60		Hexagram
5/4	$^2(5/2)$	144	36		Pentagram
4/4	$^2(4/2)$	180	0		Cross
3/4	$^2(3/2)$	240	−60		Triangle
2/4	$^2(2/2)$	360	−180		Two Points
1/4	$^2(1/2)$	720	−540		Point
0	0	∞	∞		O

Table 15: Truncated Star Polygons of four revolutions

k	k*	θ*	φ*		Truncation
∞	∞	0	180	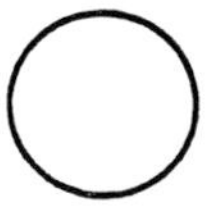	Circle
8/5	16/5	112.5	67.5		Hexadecagram
7/5	14/5	128.6	51.4		Tetradecagram
6/5	12/5	150	30		Dodecagram
5/5	10/5	180	0		Decacross
4/5	8/5	225	−45		Octagram
3/5	6/5	300	−120	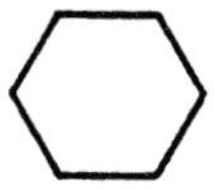	Hexagon
2/5	4/5	450	−270		Square
1/5	2/5	900	−720		Digon
0	0	∞	∞		O

Table 16: Truncated Star Polygons of five revolutions

k	k*	θ*	φ*		Truncation
∞	∞	0	180		Circle
8/6	$^2(8/3)$	135	45		Octagram
7/6	$^2(7/3)$	154.3	25.7		Heptagram
6/6	$^2(6/3)$	180	0		Hexacross
5/6	$^2(5/3)$	216	–36		Pentagram
4/6	$^2(4/3)$	270	–90		Square
3/6	$^2(3/3)$	360	–180		Three Points
2/6	$^2(2/3)$	540	–360		Digon
1/6	$^2(1/3)$	1080	–900		Point
0	0	∞	∞		O

Table 17: Truncated Star Polygons of six revolutions

k	k*	θ*	ϕ*		Truncation
∞	∞	0	180	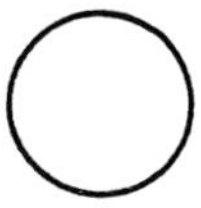	Circle
8/7	16/7	157.5	22.5		Hexadecagram
7/7	14/7	180	0		Tetradecacross
6/7	12/7	210	–30		Dodecagram
5/7	10/7	252	–72		Decagram
4/7	8/7	315	–135	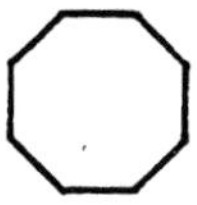	Octagon
3/7	6/7	420	–240	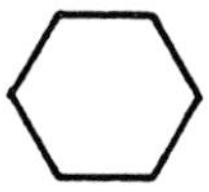	Hexagon
2/7	4/7	630	–450		Square
1/7	2/7	1260	–1080		Digon
0	0	∞	∞		O

Table 18: Truncated Star Polygons of seven revolutions

QUASITRUNCATING POLYGONS

Using the formulae, the preceding tables have predicted the results of truncating polygons with various values of k = n/d.

Whilst k remains ≥ 2 there is no problem with the physical interpretation of truncation such as was seen in Figures 19 to 22.

However when we come to the inverse polygons with k < 2, the new edge generated wants to occur outside the polygon to fulfil the angular values required. Whilst this new edge can grow outside the polygon as shown in Figure 23, its growth also causes the growth of the original edges, so that these will not consequently disappear.

Because of this the inverse polygons cannot be physically truncated and the theoretical process giving rise to the predicted polygons in the tables is called quasitruncation. This process can also be applied to the inverse star polygons as shown in Figure 24.

Quasitruncation can be interpreted in another way as shown in Figure 25. Here we regard the normal truncation process as the pulling apart of the original edges to make room for the new ones. By analogy we can with quasitruncation achieve the desired effect by pushing together the original edges and allowing the new ones to fill the gaps.

As with normal truncation, continuation of the quasitruncation process will eventually lead back to the original polygon in the form of its dual.

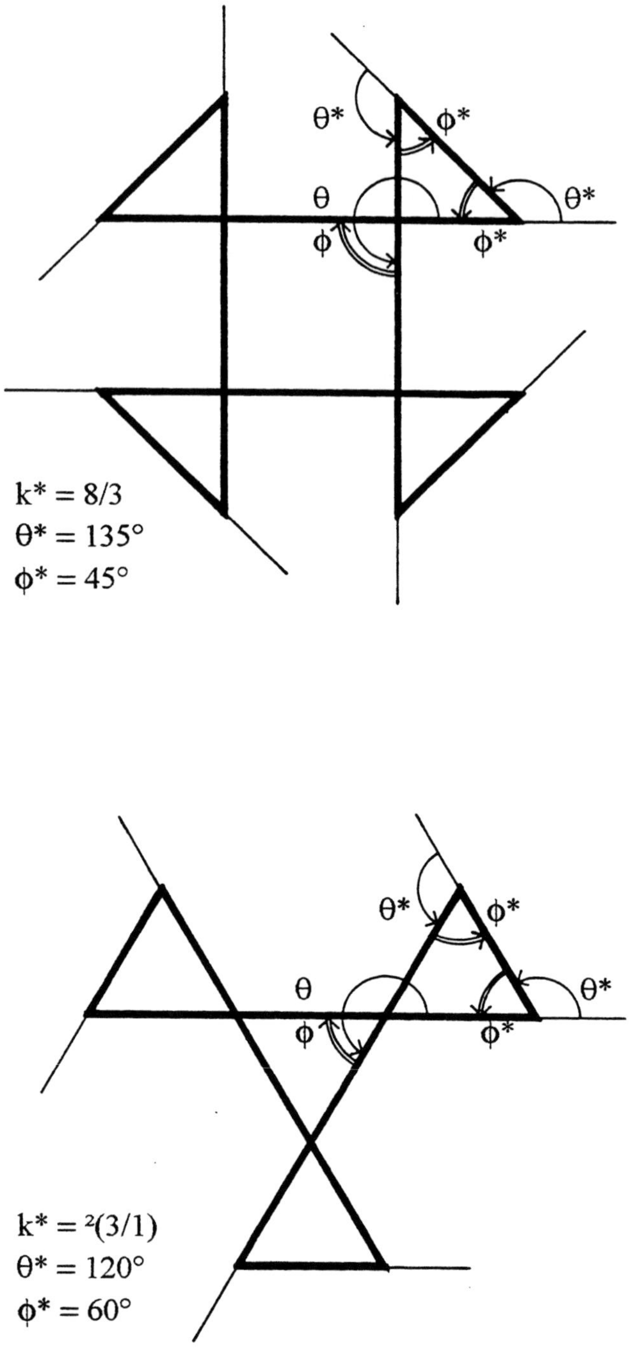

Fig.23: Quasitruncation of Polygons

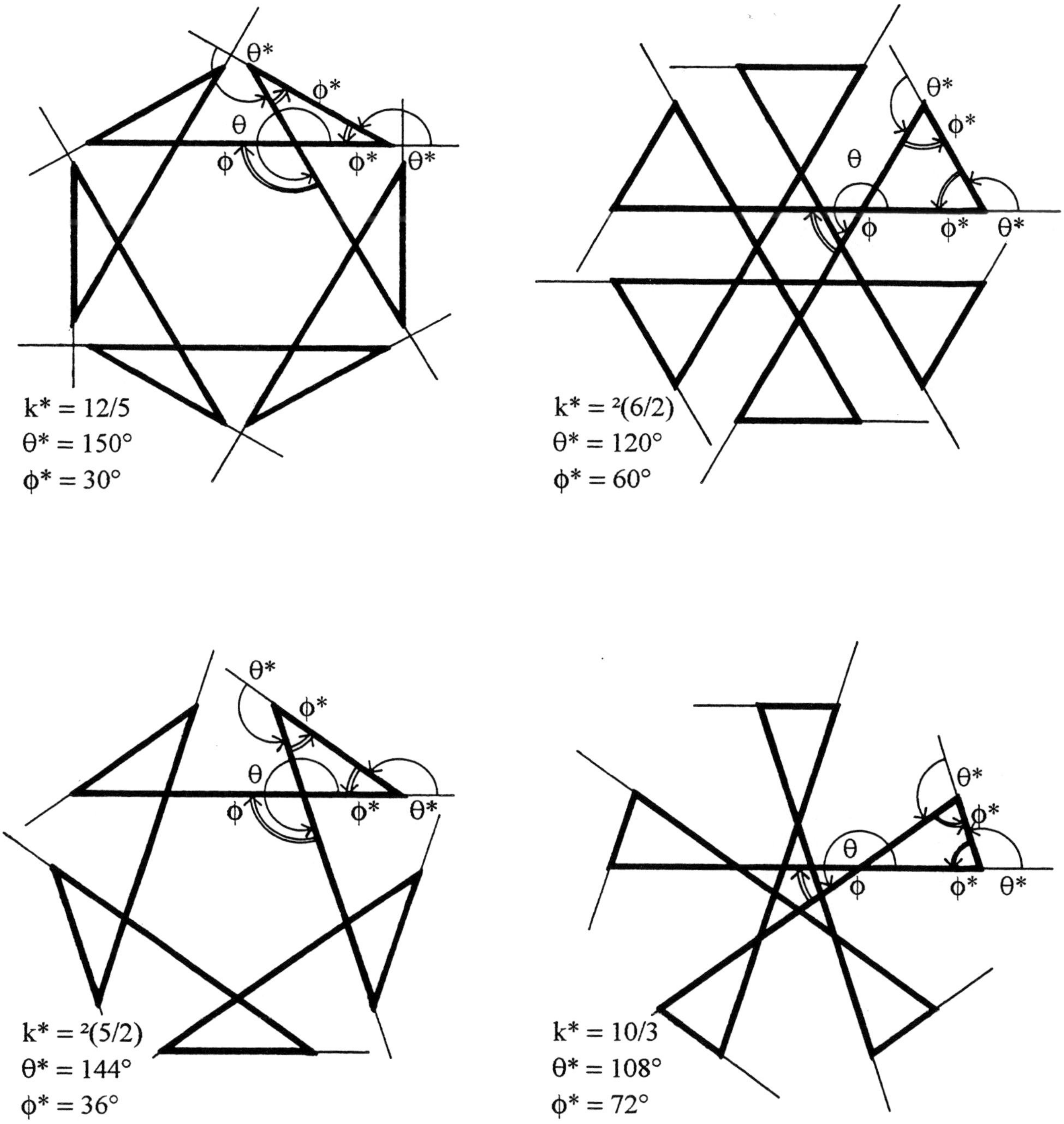

Fig.24: Quasitruncation of Star Polygons

CONJUGATE POLYGONS

The pair of polygons produced by truncating and quasitruncating an inverse pair of polygons are conjugate polygons.

The polygon defined by external angle θ and internal angle $\phi = 180-\theta$, will have an inverse where $\theta' = 360-\theta$ and $\phi' = \theta-180$. The truncation will thus have angles $\theta^* = \theta/2$ and $\phi^* = 180-\theta^* = 180-\theta/2$, whilst the quasitruncation will have angles $\theta'^* = \theta'/2 = 180-\theta/2$, $\phi'^* = 180-\theta'^* = \theta/2$, i.e. complementary internal and external angles.

Table 19 shows inverse and conjugate pairs of polygons which will be encountered later on when dealing with the polyhedra and other assemblies of polygons. As a result of their complementary angles these have related properties as was seen in Table 11. Typically the dual edge lengths of a conjugate pair can be written as $m = \sqrt{(2\pm x)}$, whilst their areas take the general form $A = n^*(\sqrt{(2\pm x)}/\sqrt{(2\mp x)})$. Here n^* for a double polygon must be taken as 2n in recognition of its double vertices thus giving an area double that apparent and in agreement with its double density.

Whilst inverse pairs are symmetrical about an axis where $d = n/2$ as was seen in Figure 10, their truncations, the conjugates, are symmetrical about an axis where $d = n/4$. This shows up with all polygons where $k = 4$ (i.e. 4/1, 8/2, 12/3 etc. or their doubles) being self-conjugate and can be seen in Figure 26 where our n vertices are again interpreted around a circle.

Inverse Pair		Conjugate Pair	
Polygon	Inverse	Truncation	Quasitruncation
2/1	2/1	4/1	4/1
3/1	3/2	6/1	$^2(3/1)$
4/1	4/3	8/1	8/3
5/1	5/4	10/1	$^2(5/2)$
6/1	6/5	12/1	12/5
7/1	7/6	14/1	$^2(7/3)$
8/1	8/7	16/1	16/7
4/2	4/2	$^2(4/1)$	$^2(4/1)$
5/2	5/3	$^2(5/1)$	10/3
6/2	6/4	$^2(6/1)$	$^2(6/2)$
7/2	7/5	$^2(7/1)$	14/5
8/2	8/6	$^2(8/1)$	$^2(8/3)$
6/3	6/3	12/3	12/3
7/3	7/4	14/3	$^2(7/2)$
8/3	8/5	16/3	16/5
8/4	8/4	$^2(8/2)$	$^2(8/2)$

Table 19: Related Polygons

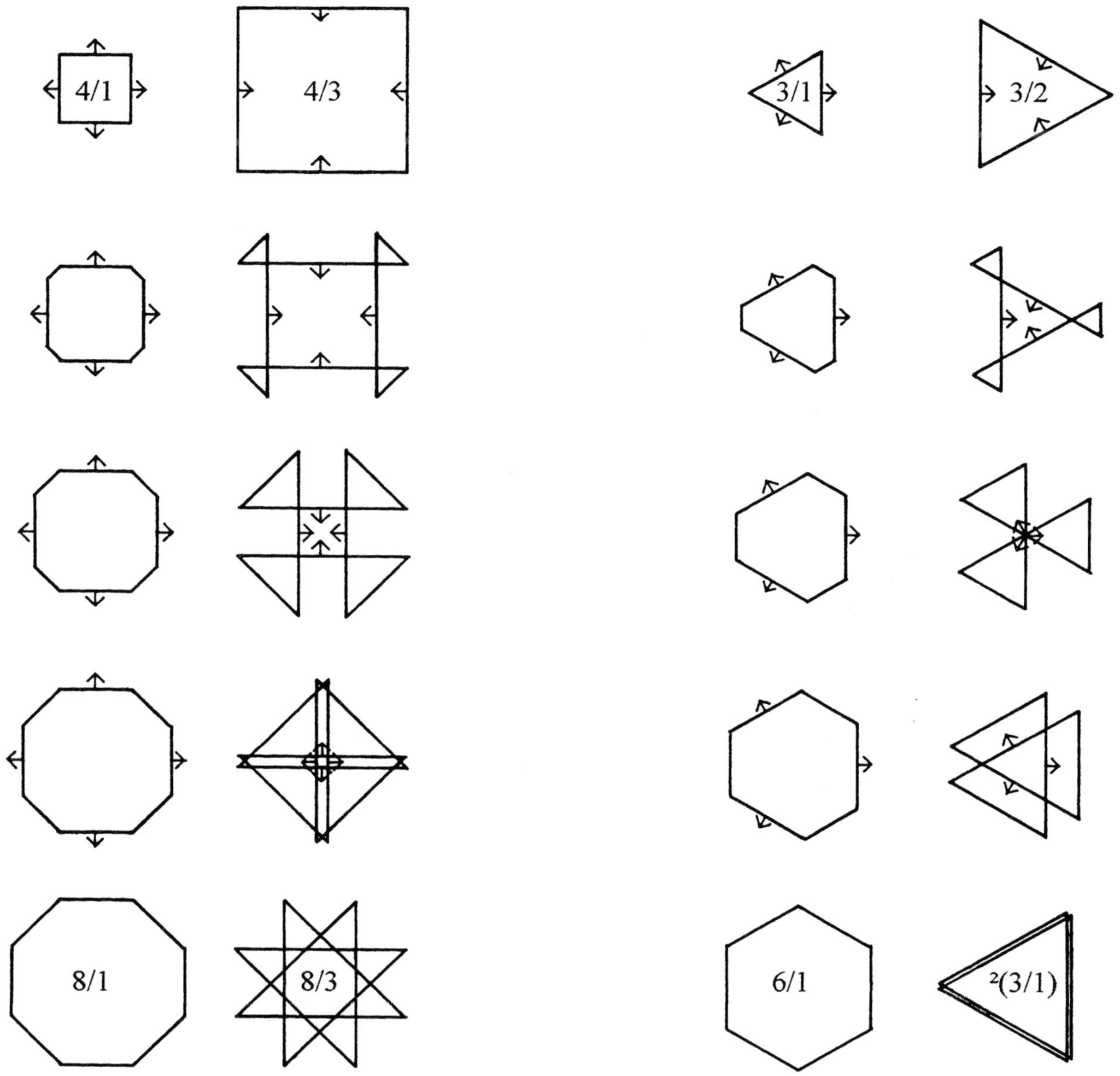

Fig.25: Interpretation of Quasitruncation of Square and Triangle

TRUNCATION GENERALLY

In the preceding tables we have covered the truncation of all the values of k = n/d that were plotted as polygons in Table 9 and it will be useful to plot these truncations in a similar way, as shown in Table 20.

What we find is that with the double polygons occupying alternate columns, the kind of pattern seen before is now more spread out.

Regular and star polygons, both single and double, now occupy a large portion of the diagram down to a diagonal line where k = 1 (k* = 2), with the even valued regular polygons with d = 1 still running off to infinity up the left hand side.

Along the line k = 1 we find single and double cross polygons, with to their right single and double, regular and star, inverse polygons, again mirroring exactly those to the left of the line.

These continue down to the line k = ½ (k* = 1) where the pattern then repeats itself again and again, firstly between k = ½ and k = ¼, with a central line of cross polygons at k = ⅓, and once more between k = ¼ and k = ⅙ with another line of cross polygons at k = ⅕ and so on.

The repetition here is the result of cycles of 360° in the value of θ* (720° in θ).

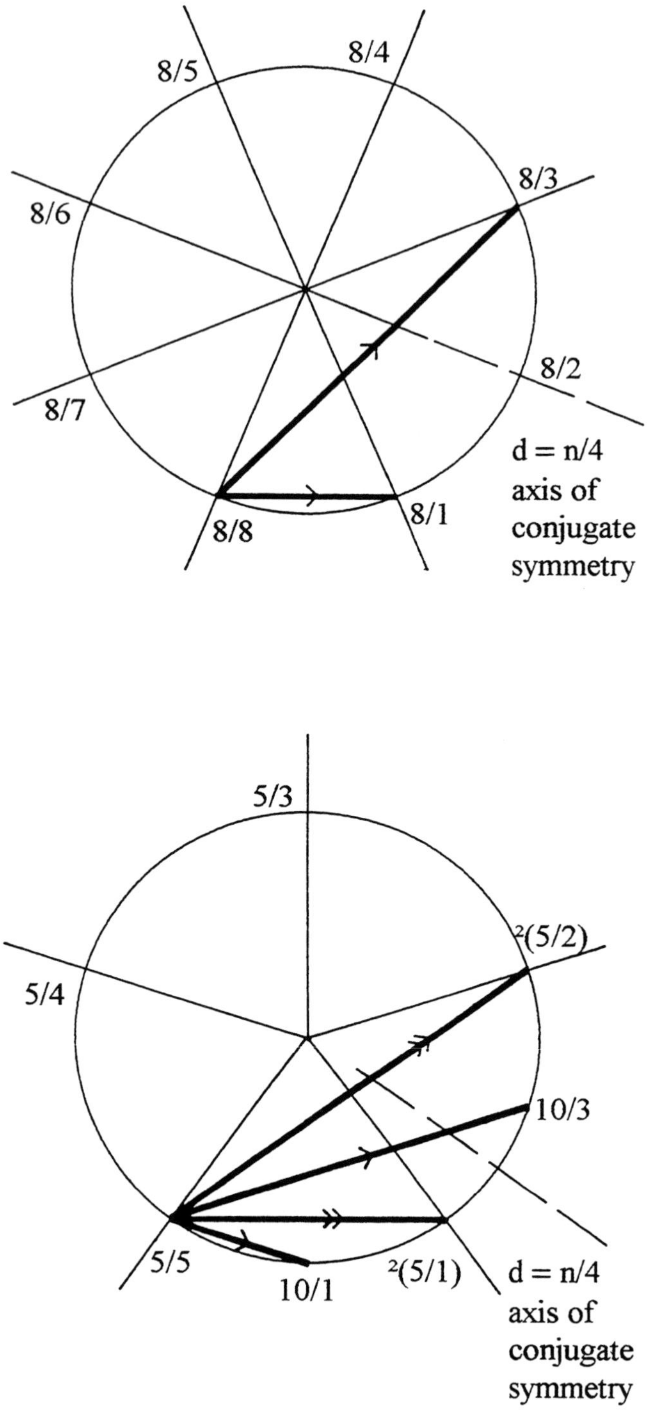

Fig.26: Interpretation around a Circle

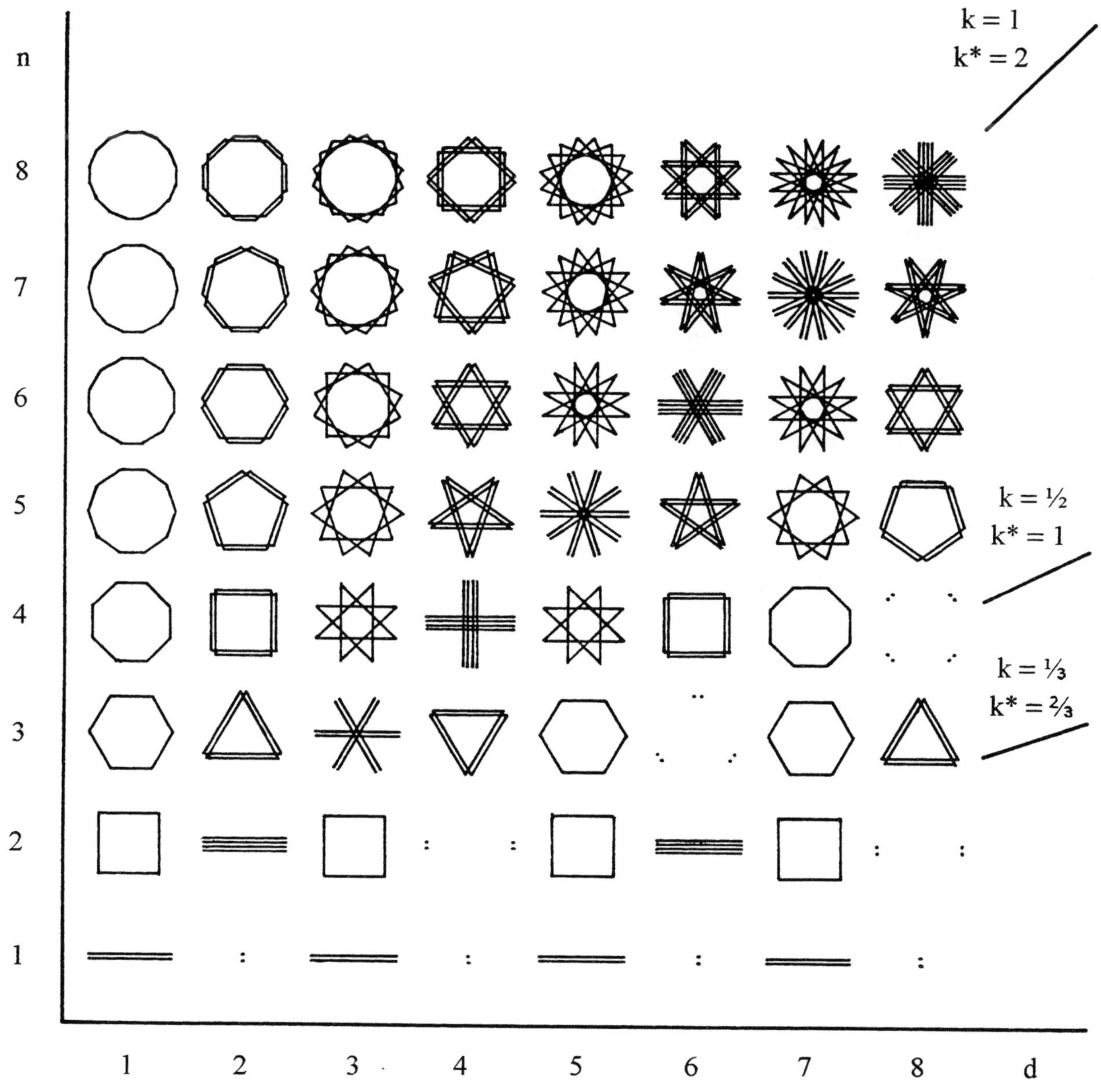

Table 20: Truncations of the Polygons k = n/d

CORRESPONDING POLYGONS

Because of this repetition, Table 20 can be simplified to only include the single set of truncated polygons, those between $k = \infty$ and $k = \frac{1}{2}$, as shown in Table 21.

Here the double polygons can be seen as forming a complete? set, exactly duplicating the pattern of single polygons in Table 10. These double polygons $k^* = {}^2(n/(d/2))$ are the result of truncating polygons $k = n/d$ with even denominators.

The remainder of the pattern consists of single polygons, but only half the original set, those with even numerators and odd denominators, the result of truncating polygons $k = n/d$ with odd denominators. By overlaying Table 21 on Table 10 we can see the result of $k^* = 2n/d$ where the integer values of n in the former correspond with the even values of n in the latter.

Returning to the even values of d, a slightly more subtle correspondence can be seen. In Table 21 any double polygon can be transformed into the corresponding polygon in Table 10 by rotating one of its two polygons through an angle $\theta^*/2$ to produce a compound polygon. This compound polygon will thus have twice as many vertices as the truncation (2n) but retain the same density (double d/2), so that once again we get 2n/d. Figure 27 shows such a pair of corresponding polygons, indeed it is the pair that Coxeter et al. unwittingly confused in their 1953 paper on Uniform Polyhedra, stating incorrectly that a 5/2 truncates to produce a 10/2!.

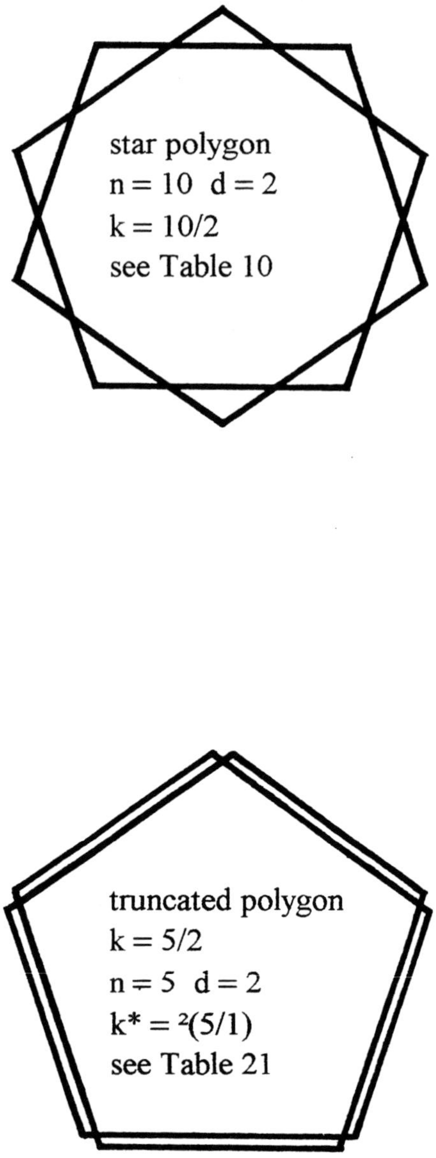

Fig.27: Corresponding Polygons

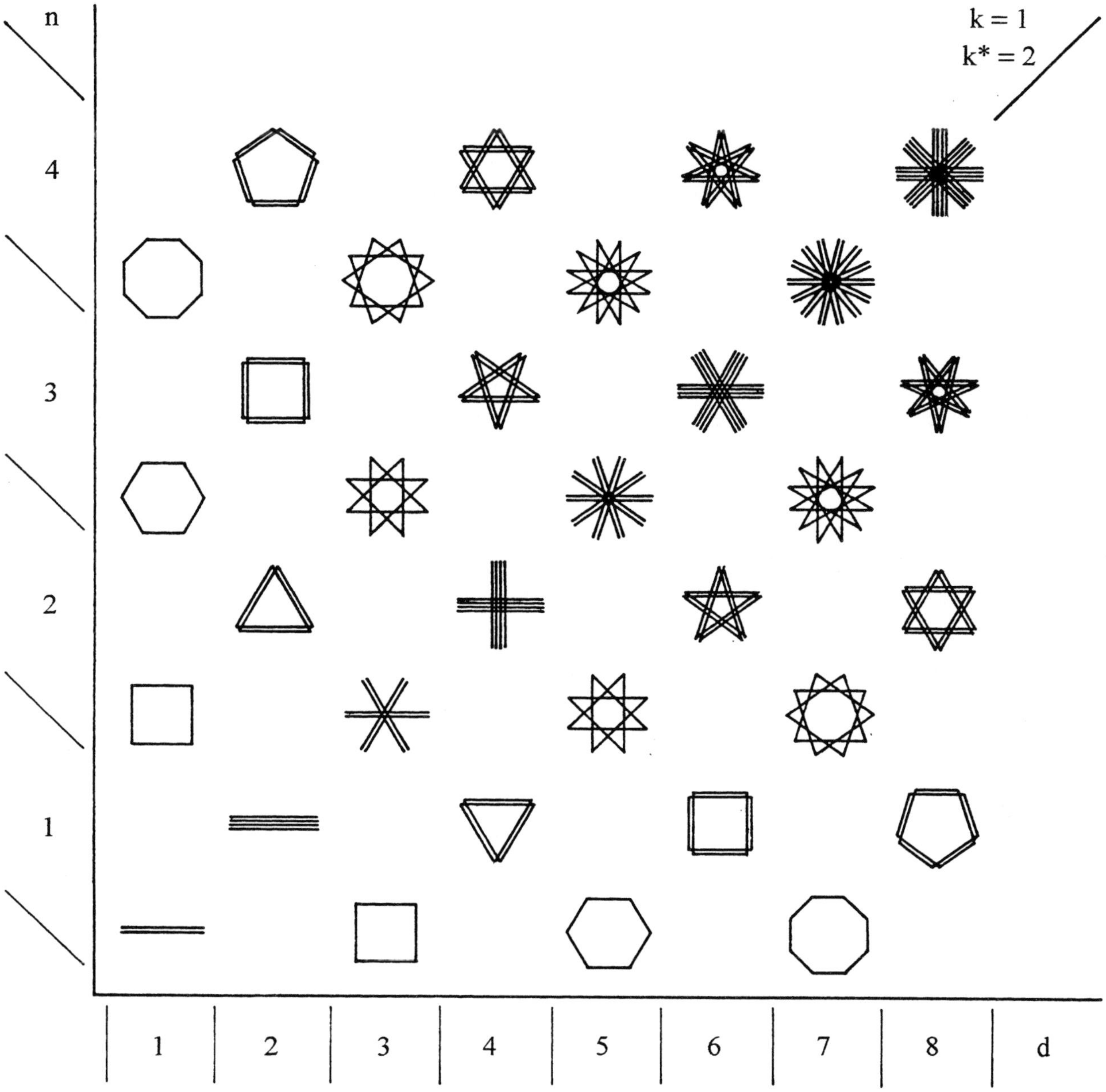

Table 21: Part of the Complete? Set of Truncated Polygons

SUMMARY OF POLYGONS

We have seen in the preceding pages that a polygon can generally be represented by a number k which is most conveniently expressed in fractional form as n/d.

The external angle of a polygon, θ was seen to equal (360/k)°, whilst the corresponding internal angle φ = (180–θ)°, and the sum of internal angles, totalled (2k–4) right angles.

By adopting a vector model to generate polygons we came to see the polygon, k = n/d, as the result of proceeding around a circle of n equidistant points, by going in turn to the dth point anti-clockwise around that circle a total of n times.
Whilst n, the numerator, came to represent the number of points one started with, d was seen to represent the number of revolutions needed to complete the figure, or equally validly the number of edges crossed to get to the centre, i.e. the density.

Inverse polygons were seen generally to proceed in the opposite direction around the circle and have the value k' = n/(n–d), such that 1/k + 1/k' = 1. Inverse pairs were seen to be symmetrical about the axis d = n/2.

Whilst n and d could have any integer values and still produce a polygon, the results were found to repeat themselves in cycles of 360° of θ, each time d reaches a multiple of n.

One complete? set, and the most simple to study, is that found for values of d in the range 1 ≤ d < n:

When d = 1, k = n/1 and a positive regular polygon results (e.g. 3/1, 4/1, 5/1, 6/1 etc)

When d > 1 a number of distinct possibilities occur:

If 1 < d < n/2 a positive star polygon results (e.g. 5/2, 6/2, 7/2, 7/3, 8/2, 8/3 etc)

If d = n/2 (for which n must be even) a self-inverse cross polygon results (e.g. 4/2, 6/3, 8/4, 10/5, 12/6 etc)

If n/2 < d < n–1 an inverse star polygon results (e.g. 5/3, 6/4, 7/4, 7/5, 8/5, 8/6 etc)

Finally in the special case when d = n–1 the result is an inverse regular polygon (e.g. 3/2, 4/3, 5/4, 6/5, 7/6, 8/7 etc)

When n and d share a common factor such that k = cx/cy, the result is a compound polygon consisting of c interlocked (x/y)s, which in the special case when x = 2y will be a cross polygon.

Regular polygons can give rise to star polygons by the process of inscribing a new polygon within or stellating, i.e. extending the edges without. The positive star polygon k = n/d produced can be seen to be the (d–1)th inscription or stellation of the original regular polygon k = n/1.

Any polygon can be truncated to produce a dual polygon of exactly the same type, that formed by joining together the mid-points of its edges. The edge length of the dual was given by the formula $m = 2\cos(\theta/2)$, a result that will be used later when forming vertex figures to describe assemblies of polygons.

The area of a polygon was given by the formula $A = n \tan(\phi/2) = n \tan(90(n-2d)/n)$. This gave a positive result for regular and star polygons, zero for the cross polygons and a negative area for the inverses. The star polygons have additional area in accord with their increasing density towards the centre.

Part way through the process of truncation an intermediate regular form is encountered, a truncated polygon with generally 2k edges. When d is odd this polygon $k^* = 2n/d$, but when d is even the result is a double polygon $k^* = {}^2(n/(d/2))$.

An inverse polygon cannot be truncated in the normal way, but by a process called quasitruncation it is possible to obtain the result predicted by the formulae for k*. This process can be interpreted as a pushing together of the edges of the original polygon creating space for new edges whilst normal truncation is seen as a pulling apart.

Conjugate polygons are those produced by truncating and quasitruncating an inverse pair of polygons. Conjugate pairs were seen to be symmetrical about the axis d = n/4.

As with the normal polygons, the truncated polygons also form a set that repeats itself, here in cycles of 720° of θ (360° in θ^*). Whilst every possible polygon can occur as a double in the truncated set, single polygons only occur with even numerators and odd denominators.

SO WHAT FOLLOWS?

It is hopefully clear from Table 10 that the set of polygons would be incomplete without the inclusion of the cross polygons. Their occurence mid-way between the ordinary polygons (both regular and star), with positive area swept out in an anticlockwise direction, and the inverse polygons with theoretically negative area, swept out clockwise, is to be expected. The cross polygons have zero area, a property often used to exclude them, and do not have a corresponding direction as all edges are diameters.

In a similar way Table 21 would be incomplete without the double polygons obtained by truncating even denominator polygons; indeed they form a more complete subset of the truncated polygons than do the truncated odd denominator polygons, which only have even numerators.

The importance of including both of these types of polygon within the overall picture will become clearer in later volumes of this series. We have not covered herein all the properties of all possible polygons, but we have hopefully laid sufficient foundations for the continuance of our study. In the same way that we have here folded lines around to enclose two dimensions within a regular polygon, we will in turn take polygons filling a two dimensional plane (tilings) and fold them up into three dimensions to enclose space within regular figures (polyhedra). These in their turn will provide the building blocks for more complex assemblies (closepacks) which can then be folded up into four dimensions (hypersolids)................

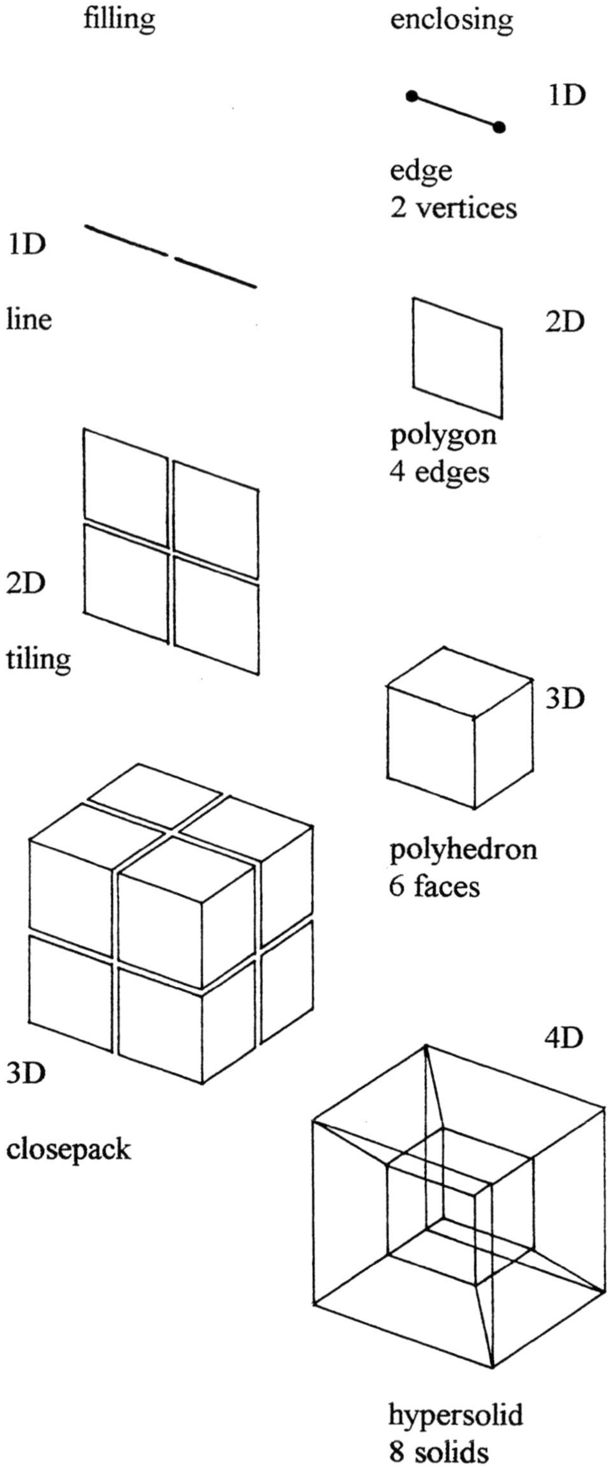

Fig.28: So What Follows?

BIBLIOGRAPHY

Abbott, E A 1962
Flatland
Blackwell

Coxeter, Longuet-Higgins & Miller 1953
Uniform Polyhedra
Phil. Trans. R. Soc. Lond. A246, p401

Coxeter, H S M 1963
Regular Polytopes
Macmillan

Critchlow, K 1969
Order in Space
Thames & Hudson

Hargittai, I & M 1994
Symmetry
Shelter

Lakatos, I 1976
Proofs and Refutations
Cambridge UP

Millington, J 1989
Curve Stitching
Tarquin

Taylor, P 1990
The Geometry Trip
Nattygrafix

Taylor, P 1995
Additions to the Uniform Polyhedra
Nattygrafix

The Open University 1981
Patterns
M101 Block 1 Unit 2

Weyl, H 1952
Symmetry
Princeton UP

ALSO PUBLISHED BY NATTYGRAFIX:

THE GEOMETRY TRIP 'A GUIDED TOUR THROUGH THE REALMS OF SYMMETRY' by Patrick Taylor 1990

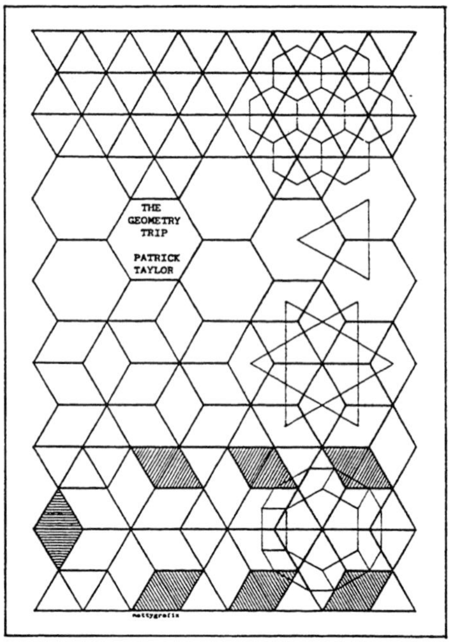

'A very thorough and dedicated work'

'should be published'

deals with the simpler polyhedra, dihedra, tesselations and skew polyhedra as varied assemblies of regular polygons

100pp A5 comb-bound including
39 line drawings and 8 tables

ISBN 0 9516701 0 7 £3.95

ADDITIONS TO THE UNIFORM POLYHEDRA by Patrick Taylor 1995

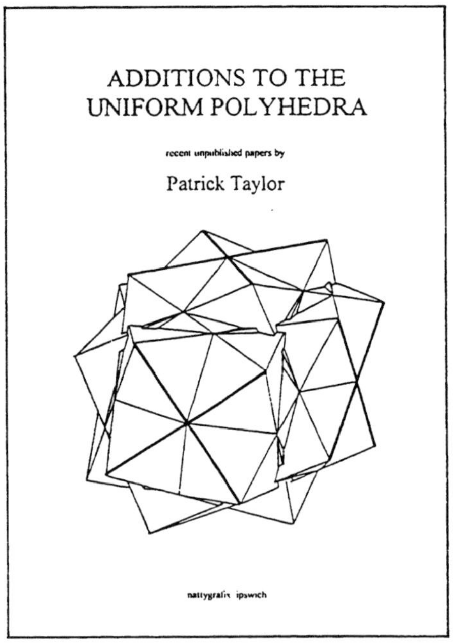

'does provide some new geometrical insights'

'extends the concept of a uniform polyhedron in an entertaining and significant manner'

presents some new configurations for inclusion within the uniform polyhedra

40pp A4 paperback including
37 line drawings and 3 tables

ISBN 0 9516701 1 5 £5.00